T0135558

Franziska Witzel

Mechanisms conferring robustness
to Erk1/2 signalling

Logos Verlag Berlin

Diss. Humboldt-Universität 2015
am Institut für Biologie

Bibliographic information published by the Deutsche Nationalbibliothek

The Deutsche Nationalbibliothek lists this publication in the Deutsche
Nationalbibliografie; detailed bibliographic data are available
in the Internet at http://dnb.d-nb.de .

© Copyright Logos Verlag Berlin GmbH 2015
All rights reserved.

ISBN 978-3-8325-4113-2

Logos Verlag Berlin GmbH
Comeniushof, Gubener Str. 47,
D-10243 Berlin
Germany

Tel.:   +49 (0)30 / 42 85 10 90
Fax: +49 (0)30 / 42 85 10 92
http://www.logos-verlag.com

Contents

1 Introduction

1.1 Important abbreviations

The following abbreviations are used to describe the different forms of Erk throughout this thesis. All other abbreviations are listed in section J.

MAPK	mitogen-activated protein kinase
Erk	extracellular signal regulated kinase, the total amount of Erk comprising isoforms Erk1 (MAPK3) and Erk2 (MAPK1)
pErk	single phosphorylated Erk; the phosphorylation lies within the activation loop, either at Thr^{202} (Thr^{185}) or Tyr^{204} (Tyr^{187}) in Erk1 (Erk2)
ppErk	dual phosphorylated Erk; both sites in the activation loop are phosphorylated
phosphoErk	is used to describe the amount of Erk detected by ppErk1/2-specific antibodies, which will include ppErk as well as some amount of pErk due to cross-specificity of antibodies [124]

1.2 The Erk signalling pathway under the influence of gene expression noise

Cells which are part of tissues and complex organisms have to behave in a concerted fashion, for which ways of communication have evolved. Cells receive stimuli from neighbouring cells or from the systemic level in the shape of diffusible growth factors. These factors bind to receptors at the cell surface and activate them. Signalling pathways consist of several proteins that together relay the signal intracellularly and elicit a stimulus-specific response. The MAPK pathway via the terminal kinase Erk integrates mainly mitogenic stimuli transmitted by growth factors and hormones [172]. Thus it is involved in fundamental cell fate decisions like cell cycle progression, proliferation, differentiation (and consequently developmental processes), migration and apoptosis [167, 121, 161].

A rough sketch of the Erk pathway is shown in Figure 1.1. After binding of growth factors like EGF (epidermal growth factor), receptors of the HER family (e.g. EGFR) form dimers, which activate each other by reciprocal tyrosine phosphorylation. In turn, a whole signalling complex crystallises around the receptor dimer. Adaptor proteins like Grb2 bind to the receptor and establish a connection with the nucleotide exchange factor Sos, which loads GTP on the small GTPase Ras, which is itself always localised to the cell membrane by a lipid anchor. Ras directly interacts with the MAPKKK (MAPK kinase kinase) Raf to mediate its localisation at the cell membrane and its activation by (de)phosphorylation events, protein-protein interactions and dimerisation [129]. Raf phosphorylates and activates the MAPKK (MAPK kinase) Mek. Mek in turn phosphorylates Erk [58]. The terminal kinase of this cascade, Erk, has hundreds of targets that reside in the cytoplasm but also within the nucleus, which are mainly transcription factors [172]. The great number of targets reflects the diversity of processes which are controlled by this signalling pathway. The best studied cytosolic targets of Erk are phospholipase A2 (cPLA2) [114] and proteins of the RSK family, the ribosomal S6 kinases [147]. By phosphorylation of S6, a subunit of the 40S eukaryotic ribosome, Erk exerts control over translation, which has to be upregulated in a concerted fashion for cells to grow and to progress in the cell cycle. Recent phosphoproteomic studies show that next to kinases, phosphatases and other signalling proteins Erk also phosphorylates apoptosis and cell cycle regulators as well as proteins like lamin A/C and Vinexin β that regulate the structure of the nucleus and the cytoskeleton [88, 32, 36]. Erk can influence nuclear transport via its substrate Nup50 (a nucleoporine) [88] and vesicle and membrane trafficking. Nuclear targets include the transcription factors c-Jun [109], JunB [36] and Elk1 [56]. The activation of transcription factors is followed by a wave of newly transcribed genes, the so-called immediate early gene products. These include EGR1 (early growth response gene 1) and c-Fos. Some of those genes act as transcription factors themselves to induce the expression of secondary growth response genes. In general, by de novo expression of a defined set of genes the cellular decision is manifested.

The aim of this thesis is to discover mechanisms that establish robustness of the MAPK signalling pathway. But what are the major perturbations that

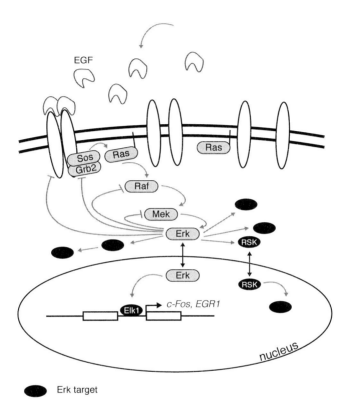

Figure 1.1 The mammalian MAPK signalling pathway. In response to growth factors like EGF growth receptors form dimers in which the monomers phosphorylate and activate each other. Adaptor proteins recruit the membrane localised small GTPase Ras, which in turn binds to Raf and mediates its activation. A kinase cascade consisting of Raf, Mek and Erk transmits the signal to cytoplasmic and nuclear targets. Members of the RSK family are bona fide targets of Erk and in this study phosphorylation of p70S6K and p90RSK have been measured to evaluate pathway activity downstream of Erk within the cytoplasm. Also, the mRNA levels of *c-Fos* and *EGR1* were measured to evaluate Erk activity within the nucleus. Inhibitory arrows indicate Erk dependent negative feedback phosphorylation.

signalling pathways have to cope with? One factor is the stochasticity of bio-chemical reactions and this issue has been addressed in detail during the last decade [15, 24, 46, 70]. Although the noise level due to post-translational mod-ification events can be relatively high, the time scale of fluctuations is usually considered too small to be transmitted to gene expression [24] and will therefore hardly influence major cellular decisions. However, increasing evidence suggests that the most important source of uncertainty in mammalian signal transduction is noise in the expression of signalling proteins. Even clonal cells display strong cell-to-cell variations of the level of the same protein with a standard-deviation of 20-30% of the mean [140]. In case of Erk, the difference in expression between

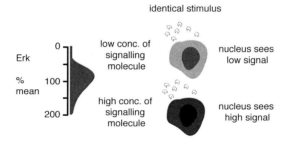

Figure 1.2   According to the distribution of Erk in single cells as illustrated in the density on the left, some cells express low levels of Erk and some cells higher levels of Erk. The same might be true for other signalling molecules. Intuitively, the amount of phosphorylated Erk is positively correlated with the total expressed amount of Erk. Therefore some cells would process stronger signals than other cells in presence of an identical stimulus.

cells in the lower and upper 10% percentile of the concentration distribution can be as high as three-fold [35]. The situation is schematically shown in Fig. 1.2. Intuitively, the amount of dual phosphorylated Erk should rise with increasing amounts of Erk. As dual phosphorylation of Erk by Mek leads to catalytic activation of Erk, this would mean that cells with higher Erk concentration would process stronger signals than cells with lower Erk expression level, when presented with the same stimulus. At the same time it is known that small quantitative differences in MAPK pathway activity can result in large changes in cellular phenotype [19, 55, 113, 133]. Erk activity has been shown to be a very sensitive gate-keeper in the G1/S transition [167] and the quantity of Erk activity is directly linked to cell-growth [93]. However, it seems unlikely that cell fate decisions are highly stochastic, especially considering tightly controlled processes like morphogenesis, where each single cell division counts.

1.3 Signalling pathways are perturbed in cancer

An even greater perturbation is imposed on signalling networks in the disease cancer. Cancer cells develop by an evolutionary process, in which they gain properties that allow them to survive and proliferate independent from systemic control of the organism. Although no tumour equals another on the genotypic level, it is believed that all tumours need to acquire a set of traits, the so-called hallmarks of cancer [62]. These are 1) self-sufficiency in growth signals, 2) insensitivity to growth-inhibitory signals, 3) evasion of apoptosis, 4) limitless replicative potential, 5) sustained angiogenesis and 6) tissue invasion and metastasis. The MAPK pathway regulates proliferation in mammalian cells. In conclusion, self-sufficiency in growth signals can be obtained by mutations of MAPK pathway components which lead to permanent pathway activity. Mutations in Ras have been identified in 30%, B-Raf mutations in about 20% of all human cancers [49].

Apart from mutations genetic alterations in cancer include gene amplifications or deletions and gene rearrangements which may lead to the aberrant expression of signalling pathway components. Frequently, tyrosine kinase receptors of the HER family are overexpressed. In colorectal cancer 4% of the samples (n=276) showed overexpression of ERBB2 [115], the prevalence was 11% (n=244) when overexpression of EGFR or ERBB2 was assessed [117]. This event is even more common in breast cancer with reported rates between 16% and 36% [16], and in non-small-cell lung carcinomas (NSCLC) where 62% out of 183 patients were found positive [69].

The mechanisms that relate increased receptor expression to increased or spontaneous pathway activity are poorly understood. The contribution to tumourigenesis is deduced from i) the obvious selection for receptor overexpression in cancer, ii) the occurring mutual exclusivity of activating receptor mutations and receptor overexpression [135], iii) the prognostic value of this event for treatment outcome [141] and iv) the positive correlation of receptor overexpression and the efficacy of therapies targeted at the receptor [130]. ErbB2 does not bind to ligands and its crystal structure suggests that this receptor is in a constitutively active conformation and thus ready to form an active dimer with another ligand-activated receptor of the ErbB family [54]. One could speculate that ErbB2 overexpression provokes an increased sensitivity of the cells to weak sub-threshold signals which would not elicit a response under normal circumstances. Also, higher expression levels might increase the rate at which the receptor monomers dimerise spontaneously.

How robust are signalling pathways towards perturbations of gene expression levels, meaning, how strong does overexpression have to be to abberantly activate a signalling pathway? Would overexpression of Erk also lead to uncontrolled pathway activity?

Though MAPK signalling is modified to be active independent of external stimuli in cancer cells, it is likely that their signalling networks preserve some robust features. This robustness often presents in the resistance of tumours to targeted inhibition of pathway components. Thus, understanding mechanisms of robustness would also help to improve therapeutic approaches in cancer.

1.4 Redundancy and systems control can establish robust features

How does the Erk signalling pathway deal with gene expression noise? The gedankenexperiment suggested that Erk activity linearly correlates with the Erk expression level (Fig. 1.2). However, it might be wrong to believe that the amplitude of the signal conveys the strength of the signal. Maybe the receiving device does not evaluate a signal in its total entity, but only certain potentially more robust features of the signal. A motivating example is the way in which grasshoppers recognise members of their own species by their characteristic songs, which is necessary when biotops are inhabited by closely related species. The spectral frequency distribution of the song from *Chorthippus biguttulus* is highly variable between single individuals [160]. Characteristics of the time dependent intensity are also fluctuating, as a song consists of 1-5 verses, where each verse consists of 20-60 syllables. However, it was found that in the face of all this variabilty, the duration of the syllable and the duration of the pause between two syllables has a constant ratio. Moreover, this quantity is robust against temperature fluctuations over a range of 24-38°C, which is because both the duration of syllable and the pause are reduced according to an exponential function at increasing temperatures. In an experiment a female grasshopper was presented an artificially generated song with changed syllable duration - when the duration of the pause was varied, individuals responded maximally at a defined pause duration [160]. In summary it was not only shown that the song (= signal) contains a robust feature, but also that this robust feature is used by the receiving system of *Chorthippus biguttulus* for species identification.

If the Erk signalling pathway has robust features, how are these robustness mechanisms implemented? A basic principle conferring stability in the light of perturbations is redundancy. If one system fails, there is another one to jump in. The circadian clock is an interesting example. In theory, to create oscillations, one gene which negatively regulates its own expression would be sufficient. However, in reality several genes and several negative feedbacks build the core clock and this makes the system robust to perturbations and more flexible regarding tissue specific expression of those genes [127, 174]. At first sight, the Erk signalling pathway seems to be characterised by high redundancy, because each kinase in the cascade exists in several isoforms. The kinase Raf exists in three different isoforms, A, B, and c-Raf, Mek and Erk appear in 2 isoforms, Mek1/2 and Erk1/2. The two isoforms of Erk indeed have the same activation kinetics and cellular function, so that the sum of Erk1 and Erk2 activity controls proliferation [92, 93]. However, in case of Raf and Mek increasing evidence suggests that isoform specific functions are present, especially on the level of activity regulation [47, 129, 33].

The concept of "systems control" explains how regulatory loops may establish robust features. The interplay of positive and negative feedback loops has been shown to underlie the robust fate decision between lysis and lysogeny for bacteria which are infected with the λ phage [99, 177]. Perfect adaptation of *E.coli* in nutrient sensing is robust towards fluctuating pathway components [2]. *E.coli* can swim towards an attractant by interrupting smooth swimming with periods of tumbling, where the bacterium sets a new direction of motion. Gradients of the nutrient concentration induce a rapid change of the tumbling frequency, however this frequency is swiftly set back to normal levels, to maintain sensitivity of the pathway. This property is called perfect adaption. Later it was shown that the network encodes integral feedback control. Here the difference of the actual system output and the desired system output is integrated over time and fed back to the system. It can be shown that the steady state always corresponds to the desired system output [171]. Another example is the Drosophila segment polarity network which is very robust against fluctuations in the kinetic properties of its components [159] and the underlying principle is bistability caused by positive feedback loops in the system [73].

To investigate systems control in the Erk signalling pathway, it is necessary to take a closer look at the multitude of regulatory features.

1.5 The regulation of Erk activity

Feedbacks shape the amplitude and duration of the pathway response

Signalling does not happen unidirectional downstream of the receptor, but also in the reverse direction by the action of various post-translational and transcriptional feedback loops. Negative feedback regulators affecting receptor ligand binding, receptor activation and early/late endocytosis of the receptor are involved in recovery of the resting state or the establishment of a new cellular state [9]. Targets of negative feedback phosphorylation by Erk are Sos [89], Raf [26, 42, 65] and Mek [28, 33] (see also the inhibitory arrows in Fig. 1.1). The main effect of these modifications is the disruption of signalling complexes: feedback phosphorylation of c-Raf by Erk impairs the interaction of Ras and Raf at the cell membrane [42] and phosphorylation of Sos dissociates the Grb2-Sos complex [89]. While constitutive tyrosine phosphatases (PTP) and serine/threonine phosphatases (like PP2A) promptly reverse phosphorylation of signalling components, the expression of some phosphatases is regulated by the activity of the various MAPKs [77, 3]. These phosphatases are called DUSPs, dual - specificity phosphatases, as they specifically remove phospho groups on threonine and tyrosine, according to the activation loop motive of MAPKs. DUSP1 additionally shows increased stability after phosphorylation by Erk [22].

The kinetics of MAPK signalling are enriched by the additional presence of positive feedback loops. Erk phosphorylates RKIP, a Raf-kinase inhibitor protein, which is supposed to elicit the dissociation of the inhibitory Raf-RKIP complex [138]. However, the *in vivo* strength of this positive feedback in comparison to the negative feedback from Erk to Raf is unknown. A very recent study found a positive feedback from Mek to Raf: phosphorylation within the N-region of c-Raf is supposed to endow c-Raf with the capability to transactivate other Raf molecules within a dimer [71]. While some phosphatases are induced or stabilised by Erk, the degradation of DUSP6 (aka MKP-3) is induced by Erk phosphorylation [102].

Aside from the aforementioned obvious regulatory loops, some regulatory loops are more subtle and result from the intrinsic kinetics of phosphorylation cycles. The first theoretical investigation that highlights such a phenomenon was the discovery that single modification cycles can produce ultrasensitive responses [57]. Later it was shown that distributive multisite (de)phosphorylation can create bistability, even without the presence of a positive feedback loop [103]. An extension is the observation that feedbacks naturally arise in cascades of phosphorylation cycles, as each layer of the cascade sequesters significant amounts of the upstream active kinase [95, 158]. For this reason it is worth to take a closer look at the mechanisms by which Erk activity is regulated at the molecular level.

Regulation of Erk activity within a dual phosphorylation cycle

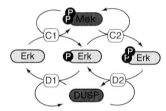

Figure 1.3 Active Mek phosphorylates Erk twice in a distributive manner. Dephosphorylation by dual-specificity phosphatases is believed to follow a distributive scheme as well. Enzyme-substrate complexes that involve the kinase Mek have been called C, complexes of Erk with DUSPs D. The appended numbers indicate first and second phosphorylation.

The regulation of Erk activity by its kinase Mek and the contribution of phosphatases is outlined in the scheme in Fig. 1.3. Erk needs to be phosphorylated on threonine and tyrosine within the TEY motif to be fully active [4]. The only enzyme that accomplishes these two phosphorylation steps is Mek1/2. *In vitro* it has been shown that Mek cannot catalyze these two phosphorylations in one reaction (as processive enzymes do), but Mek preferentially phosphorylates Erk on tyrosine first [63, 48], and then the enzyme substrate complex dissociates and reforms for the second phosphorylation step (distributive mechanism) [30]. Most likely the ATP-binding pocket of Mek1/2 is concealed when Erk is bound, so that pErk1/2 has to be released in order to load Mek1/2 with a new ATP. However, the simple fact that maybe the dissociation rate of the Mek-pErk complex exceeds the rate of the second phosphorylation step would also explain this behaviour.

Erk is dephosphorylated and thereby inactivated by different types of phosphatases. Ubiquitous phosphotyrosine phosphatases like PTP remove the phosphorylation on tyrosine. DUSPs remove phosphates on both threonine and tyrosine. Another special characteristic of DUSPs is their specific localisation either to the nucleus or cytoplasm and their regulation by MAPKs themselves. Dephosphorylation by DUSPs is believed to follow a distributive scheme as well [176].

The direct proof for distributive kinetics has been provided by *in vitro* studies. But a distributive mechanism has the potential to be converted to a quasi-processive one *in vivo*. Either molecular crowding or the anchoring to molecular scaffolds could increase the stability of the Mek-pErk complex and/or enable rapid rebinding of the latter. Up to now the experimental evidence for distributive Erk phosphorylation *in vivo* outweigh the evidence for a quasi-processive mechanism [116, 134, 153, 122]. However there is a significant controversy around this topic and the dispute will be revisited in the discussion, not least because distributive kinetics is at the basis for many interesting phenomena.

Molecular scaffolds exert spatio-temporal control in Erk signalling

The obvious players within the MAPK cascade are kinases and phosphatases, however, a number of regulatory proteins plays an essential role in determining the spatio-temporal properties of signalling. One example are scaffold proteins, which only recently came into focus of systems biology analyses [165]. Scaffold proteins bind several players of a kinase cascade and facilitate signal transmission by imposing structural input and/or allosteric regulation. Also, they have the potential to localise clusters of signalling machinery to specific locations of the cell. For example, the mammalian MAPK scaffold paxillin localises to focal adhesions [23] and is critically involved in the regulation of migration [155], the scaffold MP1 is targeted to late endosomes and enhances MAPK signalling at this compartment [166]. Specificity of MAPK signalling can be partly achieved by sequestration of Erk by cytoplasmic scaffolds like β-arrestin [137] and Sef, [154] thus preventing Erk from activating nuclear targets.

KSR (kinase suppressor of Ras), the best studied scaffold of the mammalian MAPK pathway, is constitutively bound to Mek [21] and dynamically interacts with Raf and Erk [105]. KSR resides in the cytoplasm in resting cells [173], however, just like Raf kinases, it relocates to the cell membrane after pathway

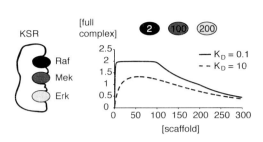

Figure 1.4 The scaffold KSR can support signalling by bringing together the three kinases of the MAPK cascade, Raf, Mek and Erk. The graph shows the concentration of the fully occupied scaffold for two different dissociation constants K_D when the single kinases are available at the concentrations as indicated on top of the graph.

stimulation. Interestingly, Erk also feedback-phosphorylates KSR, which leads to dissociation of the KSR based signalling complex [105].

1.6 Systems biology approach to investigate robustness in Erk signalling

In this thesis I analyse the robustness of Erk signalling at perturbed expression levels of Erk. The experimental part covers targeted knockdowns of Erk1/2 in colon cancer cells with constitutive MAPK pathway activation (treated in chapter 2). The complementary approach is the analysis of Erk overexpression and its impact on transient Erk signalling (treated in chapter 3). Each experimental analysis is supplemented by mechanistic mathematical modelling and was partly motivated by it.

Positive and negative feedback loops have the power to bestow robust features on signalling pathways, a concept that is called systems control. However, robust features might also arise from intrinsic properties of signalling pathways, some of which might not be obvious at first sight. Mathematical modelling is outstandingly suited for finding such properties and mechanisms. Signalling pathways can be described with mathematical models using different levels of detail and abstraction. The great benefit of this approach is the possibility to test different hypotheses prior to performing potentially expensive and time consuming experiments. In the best case the model can reveal non-intuitive behaviour that is beyond imagination when just looking at a pathway scheme of a molecular network. I want to demonstrate this with the help of an example. Imagine a scaffolding molecule enables signalling by correct positioning of the bound components and/or allosterically regulating the activity of the bound components.

It is believed that there is a clear optimum concentration at which the scaffold enhances signalling, and even small deviations from this optimum will lead to attenuated signalling [98]. The rationale behind this notion is the following. A too small concentration of the scaffold will lead to attenuated signalling, simply because effective signalling only happens with support of the scaffold. A too high concentration causes a dilution of the signalling components on different scaffold molecules — the chances for the formation of a complete signalling complex are reduced. However, considering the natural stochasticity of gene expression, it seems unlikely that the cells depend on a strictly regulated concentration of scaffold. Indeed, while the mechanism described above is intuitively understandable, it ignores a detail that changes the interpretation significantly. A simple mathematical model [168] shows that this detail is the expression level of the scaffold ligands [165]. The scaffold KSR brings together three kinases, Raf, Mek and Erk (see illustration in Fig. 1.4). In mammalian cells, Raf is typically present in a very low concentration ($\approx$13 nM in HeLa cells),[52] while Mek and Erk are rather present at concentrations of about 1 μM (also in HeLa cells) [6], thus the concentration ratio of Raf and MEK is at around 1:100. Given this difference in ligand concentration, a 100 fold change in scaffold expression would not change its ability to optimally support signalling, as i) the level of the least abundant ligand (Raf) determines the minimum amount of KSR needed and ii) the above described dilution of components on different scaffold molecules would only happen when the scaffold level exceeds the expression level of the second least abundant ligand. This is exemplified in Fig. 1.4, which shows the concentration of the full scaffold complex over a range of scaffold concentrations as predicted by the model [165].

In light of this example I formulate ordinary differential equation (ode) models that describe how Erk is activated at the molecular and/or network level. I restrict the number of described components and interactions as far as possible to create mechanistic "toy models" instead of comprehensive models that try to integrate each experimentally validated link. Some analyses are restricted to the steady state, which offers the possibility to describe pathway interactions in an indirect/abstract fashion, as described in the next section.

Modelling of signal transduction benefits from the concept of response coefficients

In the late 1990s Boris Kholodenko and coworkers have developed a mathematical concept for the quantitative analysis of signalling networks [78], which is based on the metabolic control theory. The so-called response coefficients describe how single components of a signalling network are connected with each other. The global response of a target T to a signal S is expressed as the small fractional change of T (in steady state) divided by a small fractional change in S:

$$\mathrm{R}_S^T = \lim_{\Delta S \to 0} \frac{\Delta T}{T} \bigg/ \frac{\Delta S}{S} = \frac{\mathrm{d} \ln T}{\mathrm{d} \ln S}. \tag{1.1}$$

In case of MAPK signalling, the signal S might refer to the level of epidermal growth factor (EGF), a target might be the level of phosphorylated Erk. The signal is communicated to the target via a cascade of kinases. The sensitivity of the single kinases X_i with respect to each other is characterised by the local response coefficients, defined as follows:

$$\mathrm{r}_{i-1}^i = \frac{\partial \ln X_i}{\partial \ln X_{i-1}}. \tag{1.2}$$

If a fractional change in one kinase activity leads to the same fractional change of activity in the substrate kinase, we obtain a response coefficient of 1. Local response coefficients $r > 1$ indicate signal amplification, $r < 1$ indicate signal attenuation. If the local response coefficient r_{i-1}^i is zero, it means that component X_i and X_{i-1} are not connected. However, a global response R_j^i of zero could also mean that changes in X_j are not transmitted to the level of X_i, which corresponds to the definition of robustness.

Response coefficients cannot only be used to characterise system properties like robustness, they can also serve to create models that describe the steady state of a signalling system where i) not all physical interactions are known and ii) the set of interacting components to be modelled can be chosen ad libitum [79]. The concept of response coefficients has also been successfully used for reverse-engineering cell-type specific signalling networks from pathway perturbation data [84]. I use the concept of response coefficients in this thesis to i) characterise the robustness of Erk phosphorylation with respect to Erk concentration and ii) to construct a simple negative feedback model of Erk activation.

1.7 Outline of this thesis

The first chapter is addressed to the investigation of robustness at reduced expression levels of Erk. Erk phosphorylation is shown to be partially robust towards fluctuating Erk expression levels in colon cancer cell lines. Mathematical modelling is used to test different robustness mechanisms which are subsequently investigated experimentally. The mechanism in action is identified and shown to be of relevance for successful treatment of cancer patients. Results of this study have been published [51] and I refer to this publication for the description of all experimental procedures. The second chapter focuses on robustness of MAPK signalling at Erk overexpression. I identify a novel mechanism which confers robustness to the inactive state of Erk at Erk overexpression. Motivated by this finding I have performed overexpression experiments in Hek293 cells to analyse transient signalling kinetics in response to stimulation with EGF. The main text contains a short description of all experimental methods as needed for understanding the data. Details of experimental methods and materials can be found in the appendix section. The data indicate that ppErk signal duration might be a robust pathway property. Mathematical modelling was used to propose simple mechanisms that could explain robustness of signal duration at various expression levels of Erk.

The results of each chapter are discussed separately. The conclusion consolidates results of both chapters and provides an outlook on open questions and suggestions for further experimental and theoretical investigations.

2 Colon cancer cell lines tolerate significant reduction of Erk1/2

A comprehensive version of this chapter was published in Molecular Systems Biology [51]. Raphaela Fritsche-Guenther designed all and performed most of the experiments. Additional experiments were carried out by Sandra Braun, Ricarda Herr, Nadine Lehmann and Anja Sieber. I have processed and evaluated the data, developed the mathematical models and fitted the models to the data.

2.1 Introduction

In this chapter I analyse the correlation of Erk expression and Erk phosphorylation when the level of Erk was perturbed by targeted knockdowns in various colon cancer cell lines. A description of all experimental methods can be found in the corresponding publication [51]. Some of the investigated cell lines show constitutive activity of the Erk signalling pathway due to the presence of activating mutations within either Ras or B-Raf. In that sense the measured phosphoErk levels correspond to stationary levels. Consequently, mechanistic modelling is used to suggest different mechanisms that can establish robustness of the steady state of Erk phosphorylation. For each model the response coefficient of ppErk with respect to the total amount of Erk is derived by analytical means and used as a measure of robustness. We reveal the mechanism leading to robustness in colon cancer cell lines and discuss the implications for the success of cancer therapies targeted at the kinase Mek.

2.2 Mathematical analysis predicts linear relation between protein level and activity

The activity of Erk is controlled by competition of phosphorylation and dephosphorylation of a threonine/tyrosine motive. Phosphorylation is carried out by the kinase Mek, and Erk is dephosphorylated by a multitude of phosphatases. The biochemical processes involved in Erk phosphorylation have been elucidated in depth. It has been shown that phosphorylation by Mek proceeds sequentially, tyrosine being phosphorylated before threonine [134]. Additionally, Mek tends to detach from Erk before carrying out the second phosphorylation [48], i.e. phosphorylation is not processive. Dephosphorylation is less well studied, but it is likely that it follows a similar scheme. Furthermore, it has been demonstrated that both isoforms of Erk, Erk1 and Erk2, are nearly identical in their biochemical properties [92, 172].

From this information, I developed a simple mathematical description of Erk activation where the steady state level of dual phosphorylated Erk (ppErk) is dependent on the phosphorylation rate k, the dephosphorylation rate d and the total level of Erk ($\mathrm{Erk_T}$). The differential equation system thus contains two components, singly phosphorylated pErk and dually modified ppErk

$$\frac{\mathrm{dpErk}}{\mathrm{d}t} = k \cdot \mathrm{Erk_T} - (2k+d) \cdot \mathrm{pErk} + (d-k) \cdot \mathrm{ppErk}$$
$$\frac{\mathrm{dppErk}}{\mathrm{d}t} = k \cdot \mathrm{pErk} - d \cdot \mathrm{ppErk}$$

where the occurrence of Erk has been replaced by the conservation relation $\mathrm{Erk_T}$ $-\mathrm{pErk} -\mathrm{ppErk}$. The steady state calculation yields the following expression for ppErk:

$$\mathrm{ppErk} = \mathrm{Erk_T} \cdot \frac{\left(\frac{k}{d}\right)^2}{1 + \left(\frac{k}{d}\right) + \left(\frac{k}{d}\right)^2} \ . \tag{2.1}$$

This equation shows that ppErk is predicted to be non-linearly dependent on the phosphorylation and dephosphorylation rates. The exact form of the second term is being determined by the details of the kinetic scheme, such as the number of phosphorylation sites, or whether there is cooperativity between phosphorylation of the different sites. However, the model predicts a linear dependence of ppErk on protein concentration $\mathrm{Erk_T}$, independent of the precise kinetic mechanism as long as the kinase Mek and the phosphatases are not strongly saturated.

Consequently, 30% variation in Erk levels, e.g. due to gene expression noise, would translate into 30% variation in phosphorylation of Erk in a population of clonal cells (see Figure 2.1A+B, green lines).

To quantify how ppErk depends on Erk_T in general, it is instrumental to use the normalised derivative, R:

$$R = \frac{Erk_T}{ppErk} \frac{dppErk}{dErk_T} \, . \tag{2.2}$$

This derivative corresponds to a response coefficient that defines information transfer in systems where no mass-flow exists between variables [79]. For model (2.1) R equates to one, since the phosphorylated form of the protein changes proportionally with the total protein concentration. In this case the pathway is non-robust. Values below one denote increased robustness of the phosphorylated form, and R=0 indicates perfect insensitivity of ppErk against changes in Erk levels. Examples of theoretical ppErk levels due to variations in total Erk concentration for a hypothetical non-robust, a partially robust and fully robust system are depicted in Fig. 2.1A. Taking a realistic distribution of total Erk concentrations in single cells (log-normal with s.d. of 20% of mean), a non-robust system would show strong variations in ppErk levels (Figure 2.1B), with some cells showing very low levels and other > 2 fold higher than the mean level. In contrast, a system with increased robustness (e.g. R=1/3, red lines in Figure 2.1A+B) would show a strongly reduced spread of ppErk levels. A fully robust system (R=0, blue lines in Fig. 2.1A+B) would virtually eliminate variations in ppErk levels between clonal cells.

2.3 PhosphoErk levels depend only weakly on total Erk levels

In order to test whether phosphorylated Erk is linearly dependent on the level of the protein, as the initial model suggests, one can take advantage of the fact that cells express two isoforms of Erk, Erk1 and Erk2. These isoforms are biochemically nearly identical but are expressed at different levels, with Erk2 being the dominantly expressed isoform [92]. By siRNA-mediated knock-down of each isoform alone and both isoforms together the total level of Erk was perturbed to different extents. Knock-down experiments were performed in LIM1215 cells, colorectal cancer cells without mutations upstream or within the MAPK signalling pathway [163]. Subsequently, I quantified the levels of Erk and phosphoErk from Western blots. Figure 2.1C summarises the results of this experiment. The mathematical model predicts that Erk phosphorylation is proportional to the total Erk concentration. However, the Western blot analyses clearly show that phosphoErk levels deviate strongly from the model prediction. The data rather suggest that even a reduction of Erk levels by 80% will only lead to a modest decrease of phosphorylated Erk down to 50%, thus indicating that ppErk is rather robust against variations in Erk levels. The functional correlation of Erk and ppErk can be approximated with a monomial ppErk $= \mathrm{Erk_T}^n$. In consequence, the response coefficient equals

$$R = \frac{\mathrm{Erk_T}}{\mathrm{ppErk}} \frac{\mathrm{dppErk}}{\mathrm{dErk_T}} = \frac{\mathrm{d\ln ppErk}}{\mathrm{d\ln Erk_T}} = n$$

and a linear regression of the logarithmic values leads to an estimate of R=0.43 (the monomial fit corresponding to $\mathrm{Erk_T^{0.43}}$ is shown as a solid line in Figure 2.1C).

2.4 Three possible mechanisms for robustness

Which mechanisms may lead to this remarkably low sensitivity of Erk phosphorylation towards the total level of Erk? I have focused on three properties of the signalling pathway that may lead to such low sensitivity and analysed the consequences of these potential mechanisms further utilising mathematical models.

Saturated phosphorylation by Mek

It is known that Mek is only phosphorylated at low levels [92], and that the docking interaction between Mek and Erk can be tight [87]. Therefore, Mek might be saturated by Erk, i.e. most phosphorylated Mek is bound to Erk. In that case lowering the total Erk concentration would make Mek more accessible to the remaining Erk molecules and increase the probability of their phosphorylation (depicted in Fig. 2.1D, left panel). In the following I show that already a simple model that assumes enzymatic activation of Erk by a single phosphorylation event entails robustness at the level of phosphorylation or a response coefficient

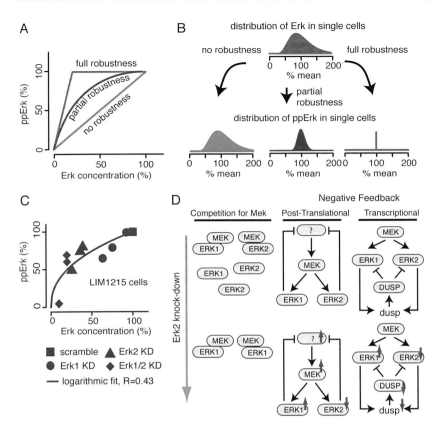

Figure 2.1 A) Mathematical analysis of Erk phosphorylation kinetics suggests that the phosphoErk level depends linearly on Erk protein concentration (green line, no robustness). The red line shows hypothetical partial robustness, where the phosphoErk level depends sub-linearly on Erk. The blue line corresponds to a fully robust system, where phosphoErk can fully compensate loss of Erk. B) The consequences of variability in Erk expression (grey) on phosphoErk expression for a non-robust, partially robust and fully robust system are shown. C) Steady state phosphoErk level of LIM1215 cells depends only weakly on Erk concentration. Each dot shows quantified pan-isoform phosphoErk and Erk levels from Western blots of cells treated with siRNA against Erk1 or Erk2 alone, or Erk1 and Erk2 in combination in percent of the scrambled control. D) Possible mechanisms providing robustness, illustrated for knock-down of Erk2: Competition for upstream kinase Mek, where loss of Erk2 results in higher access of Erk1 to Mek; Post-translational negative feedback, where loss of Erk2 results in relieve of negative feedback and therefore stronger upstream signalling; and transcriptional negative feedback, where knockdown of Erk2 results in decreasing concentrations of deactivating phosphatases.

smaller than 1, respectively. The deactivating phosphatase is assumed to work with mass-action kinetics. The steady state equation reads:

$$\text{pErk} \cdot d = \frac{v_{\max} \cdot (\text{Erk}_\text{T} - \text{pErk})}{K_\text{M} + \text{Erk}_\text{T} - \text{pErk}}, \tag{2.3}$$

where $v_{\max}$ is the maximum speed and K_M the Michaelis constant of the kinase Mek, d the rate constant of dephosphorylation. To calculate the response R one needs an expression for the derivative of pErk with respect to the total amount of Erk. I calculated the derivative of the whole equation (2.3) and obtained

$$\frac{d}{v_{\max}} \cdot \frac{\text{dpErk}}{\text{dErk}_\text{T}} = \frac{K_\text{M} \cdot \left(1 - \frac{\text{dpErk}}{\text{dErk}_\text{T}}\right)}{(K_\text{M} + \text{Erk}_\text{T} - \text{pErk})^2}. \tag{2.4}$$

Solving the equation by the derivative of interest gives

$$\frac{\text{dpErk}}{\text{dErk}_\text{T}} = \frac{\frac{K_\text{M}}{(K_\text{M} + \text{Erk}_\text{T} - \text{pErk})^2}}{\left(\frac{d}{v_{\max}} + \frac{K_\text{M}}{(K_\text{M} + \text{Erk}_\text{T} - \text{pErk})^2}\right)} < 1.$$

It is also obvious that this derivative will only assume positive values, which means that pErk increases monotonously with the total amount of Erk. Now I show that the response R is smaller than one, too. Using equation (2.4) to rewrite the response (2.2) yields

$$\text{R} = \frac{\text{Erk}_\text{T}}{\text{pErk}} \cdot \frac{v_{\max}}{d} \cdot \frac{K_\text{M} \cdot \left(1 - \frac{\text{dpErk}}{\text{dErk}_\text{T}}\right)}{(K_\text{M} + \text{Erk}_\text{T} - \text{pErk})^2}. \tag{2.5}$$

The ratio $v_{\max}/d$ in equation (2.5) can be replaced with the help of the model equation (2.3) and we arrive at

$$\begin{aligned}
\text{R} &= \frac{\text{Erk}_\text{T}}{\text{pErk}} \cdot \frac{\text{pErk} \cdot (K_\text{M} + \text{Erk}_\text{T} - \text{pErk})}{(\text{Erk}_\text{T} - \text{pErk})} \cdot \frac{K_\text{M} \cdot \left(1 - \frac{\text{dpErk}}{\text{dErk}_\text{T}}\right)}{(K_\text{M} + \text{Erk}_\text{T} - \text{pErk})^2} \\
&= \frac{\text{Erk}_\text{T} \cdot K_\text{M} \cdot \left(1 - \frac{\text{dpErk}}{\text{dErk}_\text{T}}\right)}{(\text{Erk}_\text{T} - \text{pErk})(K_\text{M} + \text{Erk}_\text{T} - \text{pErk})}.
\end{aligned} \tag{2.6}$$

The saturation of active Mek requires a high concentration of Erk that exceeds the K_M of Mek by far. Under this condition the production of pErk is limited and it can be assumed that pErk is small compared to Erk_T. The response thus simplifies to

$$R \approx \frac{\text{Erk}_T \cdot K_M \cdot \left(1 - \frac{\text{dpErk}}{\text{dErk}_T}\right)}{\text{Erk}_T \cdot K_M + \text{Erk}_T^2} < 1 \qquad (2.7)$$

which is always smaller than one as the enumerator equals $\text{Erk}_T \cdot K_M$ at maximum and the denominator is always greater than the latter product.

Post-translational feedback regulation

The MAPK-signalling pathway is regulated by post-translational feedback at many different levels. Erk has been shown to phosphorylate and thereby inactivate several adaptor molecules, and Erk deactivates c-Raf by phosphorylating inhibitory sites [40, 42]. The consequence of such negative feedback when reducing the concentration of one of the isoforms is illustrated in Fig. 2.1D. Since the amount of ppErk is reduced, feedback inhibition is relaxed, and consequently the level of ppMek increases. This in turn increases the phosphorylation of the remaining Erk isoform and the residual protein of the targeted isoform, thereby partially compensating for the loss of protein.

I will use the concept of response coefficients to (i) build a simple feedback model for the steady state correlation of Erk and ppErk and to (ii) show that the global response coefficient drops below 1 if the system contains a strong negative feedback. In the simple Erk activation model ppErk was only a function of kinetic rates and the total level of Erk (equation (2.1)). Now ppErk is also a function of the level of active Mek (ppMek). Due to the feedback loop ppMek is also a function of ppErk. So now we have:

$$\text{I} \quad \text{ppErk} = f(\text{Erk}_T, \text{ppMek}) \qquad (2.8)$$

$$\text{II} \quad \text{ppMek} = g(\text{ppErk}). \qquad (2.9)$$

Based on these equations the global response of ppErk and ppMek with respect to the total level of Erk reads:

$$\text{I} \quad \text{R}_{\text{Erk}_T}^{\text{ppErk}} = \text{r}_{\text{Erk}_T}^{\text{ppErk}} + \text{r}_{\text{ppMek}}^{\text{ppErk}} \cdot \text{R}_{\text{Erk}_T}^{\text{ppMek}} \qquad (2.10)$$

$$\text{II} \quad \text{R}_{\text{Erk}_T}^{\text{ppMek}} = \text{r}_{\text{ppErk}}^{\text{ppMek}} \cdot \text{R}_{\text{Erk}_T}^{\text{ppErk}}. \qquad (2.11)$$

The representation of the model equations with response coefficients is equivalent to a description with total derivatives, as shown in more detail in section H.1.

Plugging eq. I into eq. II and rearranging leads to the following expression for the response:

$$R_{Erk_T}^{ppErk} = \frac{r_{Erk_T}^{ppErk}}{1 - r_{ppMek}^{ppErk} \cdot r_{ppErk}^{ppMek}} .$$

As ppErk is linearly correlated with Erk_T, the local response coefficient $r_{Erk_T}^{ppErk}$ equals 1. Setting $r_{ppErk}^{ppMek} = r_1$ and $r_{ppMek}^{ppErk} = r_2$ we finally obtain:

$$R_{Erk_T}^{ppErk} = \frac{1}{1 - r_2 \cdot r_1} . \tag{2.12}$$

The local response coefficient r_1 quantifies the strength of the feedback and r_2 quantifies the signal amplification from ppMek to ppErk. Since the feedback and consequently r_1 is negative, an increased strength of the negative feedback will reduce R and thus increase robustness of Erk phosphorylation against perturbations in the total Erk level.

Transcriptional feedback

Erk is not only regulated via feedbacks at the post-translational level, but also by transcriptional negative feedback loops. So-called dual-specificity phosphatases constitute a protein family that can dephosphorylate threonine and tyrosine residues. A subfamily of these DUSPs is able to bind to Erk and is involved in Erk-dephosphorylation, mainly DUSPs 2, 4, 5, 6, 7, and 9 [120]. The expression of several of these DUSPs is controlled by transcription factors downstream of Erk thereby constituting negative feedback regulation [3, 94, 32]. Negative transcriptional feedback may provide robustness similar to post-translational feedback. If Erk concentration is reduced, expression of the inhibitory protein will be lowered and in turn the remaining Erk will be hyper-phosphorylated. Analogously to the post-translational feedback, also the transcriptional feedback needs to be strongly amplifying.

2.5 Post-translational feedback via c-Raf mediates robustness

Since each of the three mechanisms may help in compensating the changes in total Erk levels, the goal of the experimental analysis in our lab was to investigate which of them, either alone or in combination, does confer robustness to Erk phosphorylation. It was found that in general, LIM1215 cells use DUSP5,6 and 7 as negative feedback regulators. However, after Erk knockdown, none of the DUSPs1, 2, 4, 5, 6, 7, 8, 9, 10 or 16 showed significant downregulation, so that transcriptional feedback was eliminated from the list of possible mechanisms [51].

If Mek operates at saturation, it can be expected that an increased expression of Mek would revert saturation and increase the level of phosphorylated Erk. However, overexpression of Mek1 had no influence on the level of phosphoErk [51]. To investigate the hypothesis of post-translational feedback, the sensitivity of Erk phosphorylation was analysed in a panel of colon cancer cells that harbour different mutations upstream of Erk. These mutations constitutively activate the signal transduction pathway at different levels. Whereas the cell lines HT29 and RKO have an activating mutation in B-Raf (V600E), the cell lines HCT116 and SW480 contain mutations in Exon 2 of the K-Ras gene (D13E and G12V, respectively). As control, the LIM1215 cell-line is used that is devoid of KRAS or BRAF mutations. Figure 2.2A shows the position of these proteins in the MAPK signalling pathway. K-Ras activates the pathway primarily via c-Raf, whereas B-Raf activates Mek and signals independently of c-Raf [75]. If a post-translational feedback acts upstream of Ras, only LIM1215 cells, which are devoid of mutations in the pathway, should show robustness in ppErk levels. If the feedback acts at c-Raf, the expectation is that Ras-mutated cells still show robustness, but B-Raf-mutated cells would not.

I quantified the total Erk levels as well as the levels of phosphorylated Erk1 and Erk2 from these cells in which the two isoforms of Erk had been knocked down alone or in combination. As a first result, this analysis showed that there is no compensatory regulation between Erk1 and Erk2 proteins, i.e. if Erk1 protein level is lowered, this is not compensated by Erk2 protein expression and vice versa, independently of the cellular background (see Fig. 2.2B).

Strikingly, in both cell lines with a B-Raf V600E mutation, robustness is lost, i.e. the phosphorylation of Erk decreases linearly with removal of Erk (Fig. 2.2C, blue plots). In stark contrast, cells harbouring a mutation in Ras show a non-linear relation between phosphoErk and Erk (Fig. 2.2C, red plots). By regression of the logarithmic values, I estimated the global response coefficient to be R=0.36 and R=0.20 for HCT116 and SW480, respectively. This data suggests that there is an absence of robustness in B-Raf mutated cells,

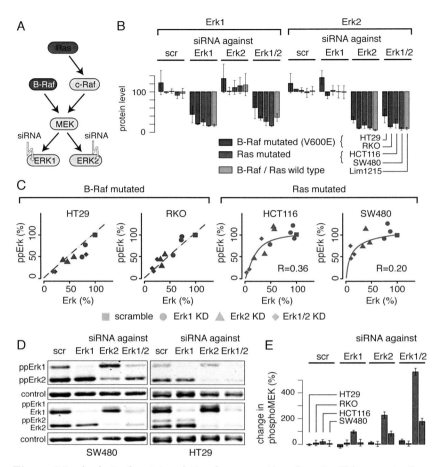

Figure 2.2 Analysis of post-translational compensation of varying Erk concentration. A) Position of mutations in the pathway: 5 colon carcinoma cell lines were analysed, LIM1215 has no mutation in the MAPK-signalling pathway, HT29 and RKO express B-Raf (V600E), and SW480 and HCT116 are K-Ras mutants. B) Changes in expression of Erk1 (left) and Erk2 (right) 48h after treating the cells with scrambled control siRNA or siRNA targeting Erk1, Erk2 or both isoforms. C) Pan-isoform Erk and phosphoErk levels after knockdown of Erk1 and/or Erk2 were calculated as fraction of the unperturbed scrambled controls. Cells with B-Raf mutation show a linear relation between Erk concentration and phosphoErk level that is predicted by a mathematical model for a system without feedback (shown as line). Cells with B-Raf wild-type show strong robustness in phosphoErk level corresponding to response coefficients of 0.36 and 0.20 for HCT116 and SW480, respectively. D) Representative Western blot images of knockdown experiments in SW480 and HT29. E) Changes in phosphoMek 48h after treatment with scrambled control siRNA or siRNA against Erk1, Erk2 or both isoforms.

whereas Ras-mutated cells or cells harbouring wild type Raf and Ras show robust compensation of loss in Erk expression. This might be due to a negative post-translational feedback targeting c-Raf, as mutated B-Raf triggers MAPK activity independently of c-Raf. In order to further investigate the hypothesis that robustness is caused by a feedback to c-Raf, we measured ppMek levels in our cell line panel after knockdown of Erk isoforms. If feedback regulation is indeed mediated by c-Raf, one would expect an increase of ppMek-levels after knockdown of the Erk isoforms in Ras-mutated cells. In contrast, one would expect no increase of ppMek in B-Raf mutated cells. As expected, ppMek levels increase in Ras-mutated cells after knockdown of Erk1, Erk2 or both isoforms in Ras-mutated cells (Fig. 2.2E, red bars). The increase in ppMek level is low following Erk1 suppression, higher after Erk2 suppression and highest following Erk1/2 double knockdown. This indicates that the effect of the feedback increases monotonically with decreased Erk protein abundance. In contrast, no increase in ppMek after Erk knockdown was observed in B-Raf mutated cells (Fig. 2.2E, blue bars). Taken together, several lines of evidence suggest that phosphoErk levels are controlled by a negative feedback, most likely at the level of c-Raf. In order to investigate whether such negative feedback sufficiently explains the observed robustness in phosphoErk level towards perturbations of the total Erk concentration, I further analysed the data using a mathematical model.

2.6 Mathematical analysis shows that the feedback is highly amplifying

The negative feedback model should describe the experimental data of ppErk and ppMek at various levels of total Erk in steady state. It was shown earlier, that in a linear model of Erk activation, ppErk linearly depends on total Erk

$$\text{ppErk} = \text{Erk}_\text{T} \cdot \frac{\left(\frac{k}{d}\right)^2}{1 + \left(\frac{k}{d}\right) + \left(\frac{k}{d}\right)^2} \; .$$

To link the level of ppErk to the level of active Mek, I integrate ppMek as a factor within the kinetic rate of activation, and replace the quotient k/d with a new (identifiable) parameter k. The model equation I (2.8) then reads:

$$\text{ppErk} = f(\text{Erk}_T, \text{ppMek})$$

$$\text{ppErk} = \text{Erk}_T \cdot \frac{(k \cdot \text{ppMek})^2}{1 + (k \cdot \text{ppMek}) + (k \cdot \text{ppMek})^2} \cdot \qquad (2.13)$$

The level of active Mek is a function of ppErk, as shown in equation (2.9). How the level changes from ppMek_0 in unperturbed cells (ppErk_0 is the ppErk level in unperturbed cells, likewise) to the new stationary ppMek level in cells with reduced amount of Erk can be approximated with a Taylor-series expansion up to the linear term:

$$\text{ppMek} = g(\text{ppErk})$$

$$\text{ppMek} = \text{ppMek}_0 + \left.\frac{\partial \text{ppMek}}{\partial \text{ppErk}}\right|_0 \cdot (\text{ppErk} - \text{ppErk}_0).$$

By rearranging the equation and substituting the partial derivative by the definition of the local response r_1 we obtain

$$\frac{\text{ppMek}}{\text{ppMek}_0} = 1 + \frac{1}{\text{ppMek}_0} \cdot r_1 \cdot \frac{\text{ppMek}_0}{\text{ppErk}_0} \cdot (\text{ppErk} - \text{ppErk}_0)$$

$$= 1 + r_1 \left(\frac{\text{ppErk}}{\text{ppErk}_0} - 1 \right). \qquad (2.14)$$

The final feedback model is built from equation (2.13) and (2.14) and contains only two parameters, k and r_1:

$$\frac{\text{ppErk}}{\text{Erk}_T} = \frac{(k \cdot \text{ppMek})^2}{1 + (k \cdot \text{ppMek}) + (k \cdot \text{ppMek})^2} \qquad (2.15)$$

$$\frac{\text{ppMek}}{\text{ppMek}_0} = 1 + r_1 \left(\frac{\text{ppErk}}{\text{ppErk}_0} - 1 \right). \qquad (2.16)$$

The parameter r_1 denotes the strength of the feedback, i.e. the relative change in phosphorylated Mek upon a relative change in phosphorylated Erk. In terms of response analysis $|r_1| < 1$ denotes weak feedback, and $|r_1| > 1$ amplifying, strong feedback. In order to investigate whether the feedback model sufficiently explains the data, I fitted this model to the measured data points for the two cell lines HCT116 and SW480 using a least-squares method. The best model fit can explain the data points well, as shown in Figure 2.3. The parameter

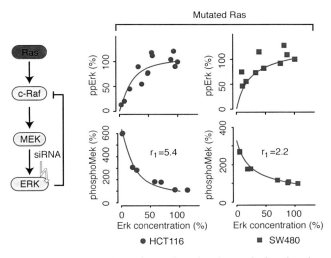

Figure 2.3 Mathematical analysis shows that the observed phosphorylation of Erk and Mek after knock-down of Erk isoforms can be fully attributed to post-translational feedback to c-Raf. Pan-isoform Erk and phosphoErk levels after knock-down of Erk1 and/or Erk2 were calculated as fraction of the unperturbed scrambled controls. A mathematical model that includes negative feedback (equations (2.15) and (2.16)) can fit the data.

r_1 was estimated to be 5.4 for HCT116 cells (k=1.5) and 2.2 for SW480 cells (k=4.0). Therefore, the model shows that the feedback from Erk to Raf is highly amplifying, where a decrease of ppErk by e.g. 10% results in approximately 20% increase of ppMek in SW480 cells, and 50% in HCT116 cells.

Taken together, the mathematical analysis shows that negative feedback from Erk to c-Raf alone is sufficient to explain the observed robustness, and the lack of robustness in B-Raf mutated cells indicates that other mechanisms are not utilised. Further experiments have confirmed that only the single feedback from Erk to c-Raf is utilised by the cells to compensate for the partial loss of Erk. Feedback phosphorylation of c-Raf at Ser^{289}, Ser^{296} and Ser^{301} was directly shown in a Western blot analysis [51]. Additionally, an activity assay of Ras confirmed that feedback regulation does not happen at the level or upstream of Ras. That no other prominent feedbacks play a role was shown with the help of Hek cells that express a fusion protein of the catalytic domain of c-Raf and the oestrogen receptor-binding domain [31]. In these cells, Raf activity can be triggered by adding 4OHT (4-hydroxytamoxifen). The fusion protein lacks the regulatory domain and is therefore insensitive to feedback regulation. When these cells were stimulated with 4OHT, ppMek was not upregulated after inhibition of Mek. Cell type specific effects in Kras and Braf mutated cells have been excluded by an analysis of feedback regulation in CaCo2 cells which expressed either Braf WT or BrafV600E. BrafV600E expressing cells lost the specific upregulation of ppMek in response to Mek inhibition [51].

2.7 Efficiency of small-molecule inhibitors is impaired by strong negative feedback

Several small-molecule compounds have been developed to target the kinase Mek, incl. the widely used experimental Mek inhibitors U0126 and AZD6244, which is currently tested in phase II clinical trials for different cancers. Both inhibitors bind non-competitively with ATP, thus reaching high specificity for Mek [170] with dissociation constants (K_D-values) in the low nanomolar range. Most likely, the inhibitors bind to both active and non-active forms of Mek. Therefore, the action of these pharmacological inhibitors can be included in model equations (2.15) and (2.16) in the following way:

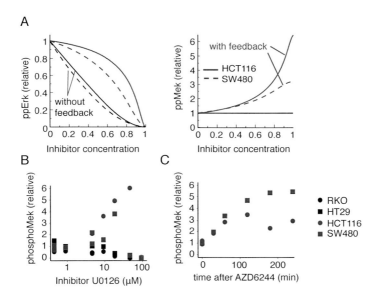

Figure 2.4 Efficiency of Mek inhibitors is strongly reduced by negative feedback to c-Raf. A) Reduction of ppErk by Mek inhibitors was simulated by a mathematical model (equations (2.17) and (2.18)) for the two cell lines SW480 and HCT116, and compared to the situation in B-Raf mutated cells, where efficiency of the inhibitors is predicted to be strongly enhanced. In B-Raf wild-type cells the model predicts a dose-dependent rise of ppMek when the inhibitor is applied. B) Experiments show that inhibition of Mek with Mek inhibitor U0126 results in strong increase in Mek phosphorylation in B-Raf wild type cell lines, while B-Raf V600E cells show no increase. C) A time series experiment after inhibition of Mek with inhibitor AZD6244 shows a rise of phosphoMek within 1 hour. At t=0, controls of untreated and DMSO-treated samples are shown.

$$\frac{\text{ppErk}}{\text{Erk}_T} = \frac{(k \cdot \text{ppMek}^a)^2}{1 + (k \cdot \text{ppMek}^a) + (k \cdot \text{ppMek}^a)^2} \tag{2.17}$$

$$\frac{\text{ppMek}^a}{\text{ppMek}_0} = (1 - I) \cdot \left(1 + r_1 \left(\frac{\text{ppErk}}{\text{ppErk}_0} - 1\right)\right) \tag{2.18}$$

where ppMek^a denotes the concentration of active ppMek, i.e. the non-inhibitor bound ppMek. The parameter I denotes the biochemical inhibitor efficiency, ppMek^I/ppMek, i.e. the fraction of ppMek that is bound by the inhibitor and thereby inactivated. Equation (2.18) is derived from eq. (2.16) with the condition ppMek= ppMek^a+ ppMek^I. Under the assumption that the inhibitor binds independent of the phosphorylation state of Mek with a low nanomolar dissociation constant, the amount of inhibitor-bound Mek equals the level of inhibitor, as described in more detail in section H.2. In consequence, the concentration of inhibitor Inh$_T$ enters equation (2.18) by $I = \text{ppMek}^I/\text{ppMek} = \text{Inh}_T/\text{ppMek}$.

Using this model, I investigated how the effectiveness of inhibitors is hampered by the presence of the strong negative feedback in B-Raf wild type cells. To describe the efficiency of Mek inhibitors in feedback-intact cells I simulate the suppression of Erk phosphorylation at varying inhibitor concentrations using the model above and the kinetic parameters that were determined for HCT116 and SW480 cells. To compare with B-Raf mutant cells the simulation was repeated with the feedback strength r_1 set to zero. The results of this analysis are shown in Fig. 2.4A. The simulations show that the efficacy of Mek inhibitors is strongly reduced through feedback-mediated robustness of the pathway. The model also shows that full inhibition of Mek will result in a strong, three to six-fold increase of ppMek levels in SW480 and HCT116 cells, respectively.

To confirm that the post-translational negative feedback via c-Raf is active in cells harbouring wild-type B-Raf upon application of a Mek inhibitor, phospho-Mek levels were measured in our lab 48 hours post inhibition of the pathway. In line with the model, phosphoMek levels rise sharply when Mek is inhibited in B-Raf wild-type cells (Fig. 2.4B). At levels higher than 20 μM, phosphoMek levels drop in all cell lines, which cannot be explained by the present model. The mechanism for this drop might be either unspecific effects of the inhibitor, inhibition of Mek phosphorylation at high doses, or indirect effects due to changed phenotype of the cells.

2.8 Discussion

Targeted knockdowns of Erk1/2 were performed in different colon cancer cell lines to analyse how the activity of one of the central signal transduction pathways, the MAPK signal transduction pathway, is influenced by changes in protein level. A linear model suggested that the active double-phosphorylated form of the terminal kinase in the pathway, Erk, would be linearly dependent on the concentration of the kinase. In stark contrast, the experiments show that the phosphorylated form is only weakly dependent on the protein concentration.

Mathematical modelling suggests that two mechanisms might account for robustness: saturation of the upstream kinase and negative feedback. The pathway is equipped with a manifold of negative feedbacks at different levels, Erk reduces the activity of various players at different levels upstream in the pathway by phosphorylation, and additionally it induces the expression of deactivating phosphatases.

It was expected that multiple feedbacks would contribute to the observed robustness, and thereby it would be hard to dissect the contribution of each feedback. The analysis of cells with different activating mutations in the pathway showed strong robustness to perturbations in Erk levels when Ras is mutated, and no robustness when B-Raf is mutated. This suggested that a single feedback from Erk to C-Raf is sufficient to explain the observed robustness. Negative feedback regulation from Erk to B-Raf that has been reported previously is either disabled by the V600E mutation, or not sufficiently strong in the K-Ras mutated cells [26, 25]. Interestingly, the post-translational feedback from Erk to c-Raf is relatively fast and acts on a time scale of one hour (see Fig. 2.4C), thus it may also confer robustness to short-term signalling. It has recently been shown that this feedback has profound effects on the robustness of this signalling pathway 20 min post stimulation, suggesting that the pathway acts as a negative feedback regulator [148].

Transcriptional feedback via phosphatases is not involved in mediating robustness. This is surprising, as intuitively one might think that variability within the pathway that changes within days may be buffered by slow feedback, which seems however not to be the case. Therefore, the role of the other feedbacks in the pathway remains puzzling. One could speculate that the post-translational feedbacks are fail-save mechanisms, i.e. that they could compensate for the loss of the feedback domain in C-Raf. Another possibility is that they actually mediate cross-talk as the other feedback targets, e.g. Sos, Src and EGFR, also activate parallel pathways. In addition, the transcriptional negative feedbacks, which are a wide-spread phenomenon [94], may be more important in fine-tuning the length of response in transient signalling to reduce noise in target gene expression [17]. Also they may play an important role in buffering strong over-activation [61].

A model predicted that targeted Mek inhibitors would be more potent in cells with B-Raf mutation than in cells without. There are several studies that investigate an association between successful inhibition of cell growth by non ATP-competitively binding Mek-inhibitors with B-Raf mutation status. A recent study showed that Mek inhibitors in B-Raf (V600E)-mutated cells show similar success as in K-Ras mutated colon cancer cells [169]. In contrast, other studies show that in cancers such as non-small cell lung cancer and thyroid cancer a B-Raf V600E is required for success of Mek-Inhibitors [50, 90, 143]. Results presented here suggest that the differences between B-Raf wild type and B-Raf V600E will be more a quantitative than a qualitative difference. The model predicts that in all cell lines that depend on MAPK signalling, Mek inhibition at higher doses will abrogate Erk signalling. However, due to loss of negative feedback, B-Raf mutated cells will respond at lower doses. In consequence, a subgroup of patients with B-Raf mutation will likely benefit from therapies targeted at Mek inhibition, in particular as side-effects towards normal cells might be minimal, as robustness of those cells will make them less sensitive to treatment.

3 Robustness of signalling against Erk overexpression in Hek293 cells

I established and performed all theoretical and experimental procedures presented in this chapter. Additional experiments (for biological replicates) were performed by Nadine Lehmann and Anja Sieber. Isoelectric Focusing was performed in collaboration with Sandra Schrötter.

3.1 Introduction

In the previous chapter it was shown how negative feedback regulation can lead to remarkable robustness of the MAPK signalling pathway in the face of varying levels of the terminal kinase Erk. The level of Erk expression was perturbed by targeted knock downs, so that only reduced levels of Erk have been considered.

A complementary approach, where the expression level of Erk is increased, is of outstanding interest as overexpression of specific proteins is a common consequence of the massive genomic alterations in cancer. In general, a higher expression of components is believed to also increase their activity. Is Erk overexpression sufficient to increase phosphoErk levels, even when the pathway is not activated, i.e. the activity of Mek is very low? In that case, Erk overexpression could replace a different activating mutation within the pathway.

Here I present a new robustness mechanism of stationary Erk phosphorylation at Erk overexpression which emerges from the distributive kinetics of Erk activation by Mek. Motivated by this finding I have performed Erk overexpression experiments in Hek293 cells to investigate pathway activity in absence and presence of mitogenic stimuli. The basal activity of the Erk signalling pathway is low in those cells, which allows the analysis of a transient signalling response.

The differential transient response in WT and Erk$^+$ cells is investigated with the help of mathematical modelling.

3.2 Mechanistic model predicts reduced Erk activity at high Erk expression levels

The mass-action model of Erk activation, which was analysed in chapter 2.2, predicts a linear correlation of total Erk and phosphorylated Erk. However, such a model cannot account for the competing docking interactions of enzymes with their respective substrates. Especially when Erk is expressed at elevated levels, saturation of the modifying enzymes can become a relevant phenomenon.

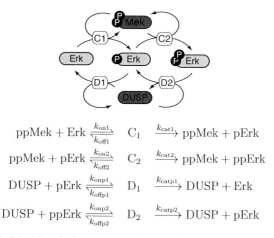

$$\text{ppMek} + \text{Erk} \underset{k_{\text{off1}}}{\overset{k_{\text{on1}}}{\rightleftharpoons}} C_1 \xrightarrow{k_{\text{cat1}}} \text{ppMek} + \text{pErk}$$

$$\text{ppMek} + \text{pErk} \underset{k_{\text{off2}}}{\overset{k_{\text{on2}}}{\rightleftharpoons}} C_2 \xrightarrow{k_{\text{cat2}}} \text{ppMek} + \text{ppErk}$$

$$\text{DUSP} + \text{pErk} \underset{k_{\text{offp1}}}{\overset{k_{\text{onp1}}}{\rightleftharpoons}} D_1 \xrightarrow{k_{\text{catp1}}} \text{DUSP} + \text{Erk}$$

$$\text{DUSP} + \text{ppErk} \underset{k_{\text{offp2}}}{\overset{k_{\text{onp2}}}{\rightleftharpoons}} D_2 \xrightarrow{k_{\text{catp2}}} \text{DUSP} + \text{pErk}$$

Figure 3.1   Model of Erk (de)phosphorylation. Active Mek phosphorylates Erk twice in a distributive manner. Dephosphorylation by dual-specificity phosphatases is believed to follow a distributive scheme as well. Enzyme-substrate complexes that involve the kinase Mek have been called C, complexes of Erk with DUSPs D. The appended numbers indicate first and second phosphorylation.

For that reason I have developed a mathematical model that satisfies the enzymatic nature of phosphorylation and dephosphorylation of Erk. Addition and removal of a phosphate group was modelled to involve the reversible formation of an enzyme-substrate complex and an irreversible step of catalysis (see reac-

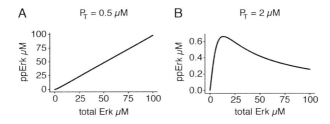

Figure 3.2 Simulation of stationary ppErk for various levels of total Erk. All parameters used can be found in table I.1. For this simulation it was assumed that the total amount of Mek is active. The amounts of phosphatase have been chosen arbitrarily and are indicated with P_T at the top of the panels.

tions in Fig. 3.1). Phosphorylation and dephosphorylation were assumed to follow a distributive scheme (see Fig. 3.1). The ordinary differential equation (ode) system associated with the presented reaction scheme can be found in the appendix section I.1. In a study on the mechanism of Erk phosphorylation in HeLa cells apparent rates of (de)phosphorylation have been measured together with the concentration of Mek and Erk [6]. I have used these parameters in my simulations and they are summarised in table I.1 in appendix section I.1.

More Erk but less Erk activation?

To determine the stationary degree of Erk phosphorylation at varying expression levels of Erk *in silico*, I have solved the ode system by numerical integration until a time point where the solution approaches an equilibrium. The steady state was confirmed with a numerical root finding routine. The approached steady state was unique and identical when the simulation was started from two opposing initial conditions, where either no Erk was phosphorylated initially, or all Erk dual phosphorylated.

The correlation of ppErk to the total amount of Erk is qualitatively different depending on the activity ratio of the modifying kinase and phosphatase. For P_T=0.5 μM (Fig. 3.2A) the maxmimum turnover rate of the kinase $v_{max,K}$ exceeds the maximum turnover rate of the phosphatase and consequently ppErk rises linearly with the total amount of Erk. However, when the condition is reversed and the phosphatase dominates with P_T=2 μM (Fig. 3.2B), the correlation of Erk and ppErk becomes nonlinear, where ppErk at first increases,

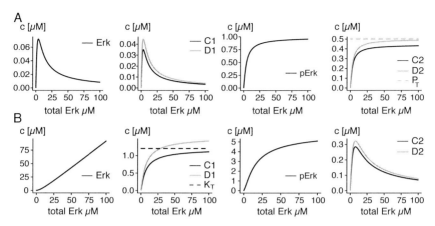

Figure 3.3 The dual phosphorylation cycle in steady state. The stationary amount of all components in a dual phosphorylation cycle when the total phosphatase level is A) P_T=0.5 μM or B) P_T=2 μM. All other parameters like in Figure 3.2. Dashed lines indicate the total level of the kinase or phosphatase as referenced by the legend.

but starts to decrease for higher levels of Erk, eventually approaching zero as its limit.

The analysis shows that as long as the phosphatases dominate or in other words, as long as the kinase activity is low, the formation of excessive amounts of active Erk is suppressed. It can be expected that this condition is fulfilled in unstimulated mammalian cells, as they regulate pathway activity via the activity of the kinase whereas effective phosphatases are abundant at any time.

What is the mechanism behind the non-linear correlation of Erk expression and Erk activation? The single phosphorylated Erk increases monotonically with the Erk expression level, however, it also approaches a limit (Fig. 3.3B, 2nd panel from the right). The ppMek-Erk enzyme-substrate complex C_1 behaves accordingly, with the limiting concentration given by the total amount of ppMek, here called K_T (Fig. 3.3B, 2nd panel from the left). Obviously ppMek is saturated with unphosphorylated Erk at increasing levels of the latter. Limited activation with exactly the same characteristics has been described for a single modification cycle previously [96]. However, there is a more intuitive approach to understanding the activation limit in a single phosphorylation cycle, which I will show in the next section.

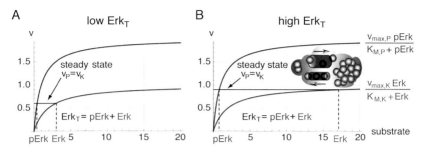

Figure 3.4 Overexpression insensitivity in a single phosphorylation cycle. A) The velocity of the kinase (pink) and the velocity of the phosphatase (blue) as a function of substrate level according to Michaelis-Menten. The steady state levels of Erk and pErk are indicated with the dashed lines. B) For higher levels of Erk enzyme velocity is increasing with increasing levels of the substrate, however the maximum speed of the cycle is determined by the kinase, as $v_{max,K} < v_{max,P}$. Kinase and phosphatase can be seen like pipes with different diameter, where the diameter of the pipe sets the maximal flow rate. The steady state condition demands the same flow rate for these two pipes (3 molecules per pipe per time unit in this sketch). When the kinase is saturated its speed is no longer proportional to the amount of substrate (zero order kinetics) and unphosphorylated Erk accumulates, while the phosphatase operates in the linear regime.

Limited activation in a single phosphorylation cycle

Both enzymes, the modifying kinase and phosphatase, have a rate of catalysis that is shaped by affinity to the substrate K_M and the turnover rate v_{max} as shown in Figure 3.4A. For each enzyme alone it holds that the rate is increasing with the amount of substrate until the enzyme saturates. In steady state both enzymes work with the same turnover rate and the ratio of Erk to pErk follows as indicated by the dashed lines in Fig. 3.4A. In the cells resting state the phosphatase has a higher maximum speed than the kinase, as illustrated in Figure 3.4. Due to the steady state condition of equal rates for the kinase and phosphatase, $v_{max,K}$ is the maximum speed at which the cycle can run when the level of Erk_T is increased even further. In consequence only Mek saturates and the amount of phosphorylated Erk has an absolute limit, while unphosphorylated Erk accumulates (see Fig. 3.4B).

The steady state of a single phosphorylation cycle has an analytical solution [57] and the maximal amount of the activated species when $v_{max,K} < v_{max,P}$ can be derived by calculating the limit for large values of the target [96].

However, given the description above, the calculation can be simplified. As were are looking at the limit for high substrate levels, the Michaelis-Menten model is an adequate description. The assumption that the amount of substrate bound in an enzyme-substrate complex is small compared to the free substrate is fulfilled. Now, following the illustration in Fig. 3.4 the maximum activation level can be derived from the steady state condition:

$$v_P = \frac{v_{max,P} \cdot pErk_{max}}{K_{M,P} + pErk_{max}} = v_{max,K} \leftrightarrow$$

$$pErk_{max} = \frac{K_{M,P}}{\frac{v_{max,P}}{v_{max,K}} - 1}. \tag{3.1}$$

The activity ratio of kinase and phosphatase directly influences the stationary level of phosphorylated Erk. When the maximal turnover rate of the phosphatase is twice the maximal turnover rate of the kinase, the amount of phosphorylated Erk complys to the Michaelis-Menten constant of the phosphatase. The role of the phosphatases' K_M is intuitive, as a weaker affinity of the phosphatase helps to pile up more of the activated species pErk.

In conclusion, activation of a protein already within a single modification cycle confers insensitivity to overexpression of the protein, when the activating enzyme has a smaller activity than the deactivating enzyme, as shown previously [96]. That means i) the mechanism is effective, when the cells are in a resting state with no/low kinase activation and ii) overexpression of Erk alone would not lead to an equivalent increase in Erk activity, if Erk would only be activated by a single phosphorylation event. The appendix section I.2 additionally contains an elegant derivation of the Erk concentration which leads to the half-maximal amount of pErk.

The signal is attenuated further in a dual phosphorylation cycle

The excursion to single phosphorylation cycles helps to understand the dual phosphorylation cycle. Here the stationary level of pErk rises with the total amount of Erk, but only up to a certain limit, just like in a single phosphorylation cycle. Also the condition for this steady state scenario is a higher maximal turnover rate of the phosphatase compared to the kinase. Regarding saturation of ppMek, in the given system ppMek might either be free or it can be in

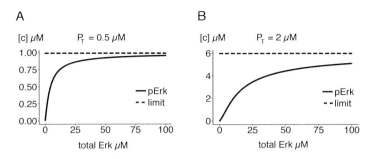

Figure 3.5 The limit of single phosphorylated Erk in a dual phosphorylation cycle can be calculated analytically and is indicated by the dashed line. The total phosphatase level is A) P_T=0.5 μM and B) P_T=2 μM. All other parameters for the simulation of stationary pErk were set like in Figure 3.2.

a complex with either Erk (C_1) or pErk (C_2). As ppMek gets saturated, the amount of free kinase goes to zero - given the continuous increase of pErk and the non-monotonic behavior of ppErk, it seems like with increasing amounts of Erk more and more of the kinase is sequestered in the complex C_1. That means, two mechanisms shape the steady state amount of ppErk at Erk overexpression: saturation of ppMek and sequestration of ppMek in the first phosphorylation step. As a consequence, also the phosphatase is drawn into the first phosphory-lation cycle - D_2 and C_2 decrease for rising levels of total Erk (Fig. 3.3B, right panel).

When the condition is reversed, so when $v_{\text{max,K}} > v_{\text{max,P}}$, all intermediate species of the dual phosphorylation cycle behave in a mirror-inverted fashion, e.g. unphosphorylated Erk exchanges its concentration profile with the profile of dual phosphorylated Erk. The phosphatase saturates in the 2nd phosphorylation cycle and draws most of the kinase into the 2nd cycle (Fig. 3.3A).

As the kinase (phosphatase) is sequestered in the first (second) phosphory-lation cycle, the limit of single phosphorylated Erk in a dual phosphorylation cycle can be calculated like in a single phosphorylation cycle:

$$
\text{pErk}_{\max} = \begin{cases} \dfrac{K_{\text{M,P1}}}{\frac{v_{\text{max,P1}}}{v_{\text{max,K1}}}-1} & \text{when } v_{\text{max,K}} < v_{\text{max,P}} \\[3mm] \dfrac{K_{\text{M,K2}}}{\frac{v_{\text{max,K2}}}{v_{\text{max,P2}}}-1} & \text{when } v_{\text{max,K}} > v_{\text{max,P}} \ . \end{cases}
\tag{3.2}
$$

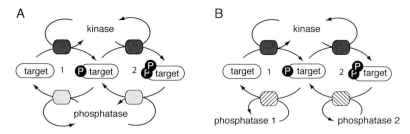

Figure 3.6 Different phosphatases might catalyse dephosphorylation of Erk. A) Dual-specificity phosphatases remove phosphates from Thr and Tyr. B) Alternatively, abundant Tyr Phosphatases and Thr/Ser phosphatases might work in parallel.

How the equations in (3.2) can be derived is exemplary shown in appendix section I.2. $K_{M,P1}$ refers to the affinity of the phosphatase in cycle 1, which is the affinity with respect to unphosphorylated Erk. Likewise $K_{M,K2}$ refers to the affinity of the kinase in cycle 2, so with respect to pErk. Maximum turnover rates v_{max} are labelled accordingly. Figure 3.5 shows the amount of single phosphorylated Erk in a dual phosphorylation cycle for increasing amounts of Erk and the calculated limits using the equations (3.2).

Different phosphatases are involved in Erk deactivation

The maximum turnover rate of an enzyme equals the product of the catalytic rate and the concentration of the enzyme. Here one enzyme is responsible for (de)phosphorylation of Thr and Tyr, so that it is unlikely that the condition $v_{max,K} > v_{max,P}$ is fulfilled in the first cycle but not in the other and vice versa. But dual-specificity phosphatases are a class of phosphatases whose expression is highly regulated in concentration and location [74]. Under some circumstances they might not even be the main phosphatases responsible for deactivation of Erk. In the scenario, where dephosphorylation of Thr and Tyr is carried out by two different phosphatases, the activity ratio of kinase and phosphatase can differ in the two cycles. The two different modes of Erk deactivation are shown in Fig. 3.6. When the phosphatase outcompetes the kinase in both cycles, ppErk shows the same non-linear profile as was seen before (Fig. 3.7A). The formation of ppErk is suppressed in the same way when the phosphatase has a greater turnover rate only in the first phosphorylation cycle (Fig. 3.7B). In contrast,

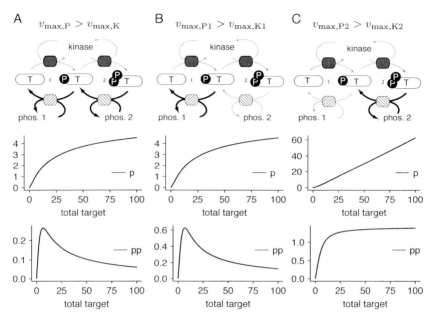

Figure 3.7 Limited activation in dual phosphorylation cycles in general. The figure shows the stationary amount of single (p) and dual phosphorylated target (pp), when two different phosphatases deactivate the target. In A) v_{max} of the phosphatase exceeds the level of v_{max} of the kinase in both cycles. In B) and C) the phosphatase has a higher turnover rate only in one out of the two cycles as indicated at the top of the panels.

when the kinase dominates in the first phosphorylation cycle, the formation of single phosphorylated Erk is unlimited. However when the phosphatase has a higher turnover rate in the 2nd cycle, the kinase saturates and ppErk approaches a limit, just like pErk would in a single phosphorylation cycle (Fig. 3.7C).

It can be concluded that as long as the phosphatase dominates the activity of the kinase in at least one cycle, activation of Erk is limited even at higher expression levels.

A simplified model explains limited activation in a dual phosphorylation cycle

Based on the analysis presented in this chapter it is possible to reduce the complexity of the model. Limited activation of Erk is seen when ppMek is shared between two cycles and eventually saturates and sequesters in one of the cycles. The phosphatases keep working far from saturation, so that it can be assumed that they proceed with mass-action kinetics instead. Even with this modification, the explicit description of all components in steady state is impossible, which is generally true when the various enzyme-substrate complexes are appreciable compared to the concentration of free substrate and product [132]. However, I can approximate

$$\mathrm{K} \approx \mathrm{K_T} - \mathrm{C_1} \tag{3.3}$$

$$\mathrm{Erk} \approx \mathrm{Erk_T} - \mathrm{C_1} - \mathrm{pErk} \tag{3.4}$$

because the concentration of the complex formed by ppMek and monophosphorylated Erk, $\mathrm{C_2}$, is significantly smaller than $\mathrm{C_1}$ and ppErk has the smallest contribution to the total level of Erk.

In equation (3.3) and (3.4) $\mathrm{Erk_T}$ and $\mathrm{K_T}$ denote the respective total enzyme concentrations of Erk and ppMek. The steady state of this simplified system has a closed form and can be found in appendix section I.3. The approximation of the steady state captures the correlation of phosphorylated Erk and total Erk qualitatively as well as the order of magnitude in phosphorylation, as can be seen in a direct comparison of the numerical solution of the system with mass-action kinetics for dephosphorylation with the analytical approximation (Fig. 3.8) where the conservation relations of Erk and ppMek have been truncated as shown in equation (3.3) and (3.4).

Most informative about the system is the concentration of Erk, at which maximal amounts of ppErk are formed. The derivative of ppErk by the level of total Erk

$$\frac{\mathrm{d}}{\mathrm{dErk_T}} \mathrm{ppErk}\,(\mathrm{Erk_T}) = \gamma \frac{\mathrm{dC_1}}{\mathrm{dErk_T}} [\mathrm{K_T} - 2\mathrm{C_1}(\mathrm{Erk_T})] \overset{!}{=} 0 \tag{3.5}$$

equals zero at the maximum (γ is a constant factor, for derivation of eq. (3.5) see section I.3). Condition (3.5) is only fulfilled when

$$\mathrm{C_1}(\mathrm{Erk_T}) = \frac{\mathrm{K_T}}{2}. \tag{3.6}$$

The level of total Erk in the cell leading to maximal activation is the one where half of the total available kinase ppMek is sequestered in a complex with unphosphorylated Erk. That means, most Erk can be activated, when equal amounts of the kinase are available for the first and the second phosphorylation step. Condition (3.6) allows for the exact calculation of the maximum coordinate to

$$(\mathrm{Erk_T}, \mathrm{ppErk})_{\max} = \left(K_{\mathrm{M1}} + \left[1 + \frac{k_{\mathrm{cat1}}}{d_1} \right] \frac{\mathrm{K_T}}{2}, \frac{k_{\mathrm{cat1}}k_{\mathrm{cat2}}}{d_1 d_2 K_{\mathrm{M2}}} \cdot \frac{\mathrm{K_T}^2}{4} \right). \qquad (3.7)$$

Position as well as height of maximum Erk activation are biological significant properties. The position limits the concentration domain in which the cell reacts with increased activation to an increased level of protein. The Erk concentration which leads to maximal ppErk level is directly proportional to the concentration of ppMek. The maximal ppErk level is proportional to the square of the kinase concentration which reflects the two step nature of the activation process. A higher affinity of the kinase to nonphosphorylated Erk (smaller K_{M1}) enforces sequestration and thus shifts the position of the peak to smaller levels of Erk. A higher affinity in catalysis of the 2nd phosphorylation (smaller K_{M2}) increases the activation level. Only the catalytic rates of the 1st modification cycle (d_1 and k_{cat1}) influence the peak position, which confirms that the activity ratio of kinase and phosphatase in the cycle converting between Erk and pErk creates the prerequisite for limited activation.

Quantification of the activation limit

With the kinetic parameters that have been measured for HeLa cells and assuming that only 5% of the cellular Mek is activated, maximal levels of active Erk can be found at 2.3 μM which is about 3 fold more than the average Erk expression level measured in HeLa cells (Fig. 3.8). Also, only 2% of Erk is activated at the peak, which means that 5% Mek activity is attenuated to only 2% of Erk activation at the peak. For the physiological concentration of Erk, at 0.74 μM, indicated with the dashed vertical line in Fig. 3.8, the relative Erk activation is at 4.5%. Single phosphorylated Erk approaches a limit, which accords to 0.67 μM (calculated from equation (I.15) in section I.3). The half maximal amount of pErk is found at an expression level of 2.7 μM, where 12% of Erk is single phosphorylated. This also corresponds to an attenuation of phosphorylation compared to 19% of single phosphorylated Erk at the physiological

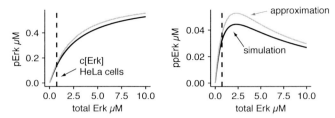

Figure 3.8   Analytical approximation of the simplified dual phosphorylation model. The steady state level of pErk and ppErk at various levels of Erk_T was simulated with the model that features distributive dual phosphorylation of Erk by Mek and mass action rates of dephosphorylation (see eq. system (I.6)). The conservation relation for K_T and Erk_T is either complete (simulation, black line) or truncated according to eq. (3.3) and (3.4) (approximation, grey line). The dashed line indicates the concentration of Erk in HeLa cells [6].

average expression level of 0.74 μM (all of those numbers refer to 5% activation of Mek). With the help of the analytic formulas derived here, the maximal activation level of a target can be estimated for any single or dual phosphorylation cycle, given that the catalytic rates are known. In case of Erk activation in HeLa cells, the mechanism which limits Erk activation is effective already at 3x overexpression, which can be considered mild in comparison to the observation that Erk concentrations vary about 3 fold easily between different cells.

An intuitive description of the mechanism

I preclude this chapter with an intuitive description of the phenomenon. At high expression levels of Erk, the single molecules are standing in a waiting line for a ticket, which is a phosphate group that they obtain from ppMek (Figure 3.9, step 1). Every time an Erk molecule gets a ticket, it has to go back to the end of the waiting line and wait for the second ticket (step 2). Phosphatases are abundant and may steal the ticket anytime (step 3). Now, the more Erk there is, the longer the waiting time for the second ticket, and the higher the probability that the first ticket is lost, before the second ticket can be obtained. This analogy makes it intuitive, why i) more Erk leads to less dual phosphorylated Erk and ii) why the activity ratio of phosphatase and Mek in the first phosphorylation step is critical for the mechanism.

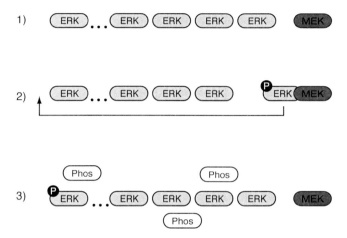

Figure 3.9 Overexpression insensitivity in a dual phosphorylation cycle: a queuing situation. If Erk is expressed at much higher levels than its kinase Mek, availability of Mek becomes the limiting factor. A waiting line is an excellent metaphor for this situation.
Step 1) Erk molecules are standing in a waiting line to obtain their first phosphorylation from Mek.
Step 2) As the dual phosphorylation is distributive, pErk has to go to the end of the queue and wait for the second modification.
Step 3) While pErk advances in the queue, it may encounter one of the abundant phosphatases and loose its first phosphorylation before the second one can even occur. The higher the Erk expression level, the longer the queue and the more unlikely it is that Erk obtains both phosphorylations which are required for full activity. In the reverse situation where Mek has a higher catalytic activity then the counteracting phosphatases, Mek and Phos reverse roles in this setup.

3.3 Hek293 cells as experimental model for Erk overexpression

Cloning of Erk2 expression constructs

How is cell signalling affected when the kinase Erk is overexpressed? To tackle this question experimentally, the first step along the line was to design and clone a suitable expression construct. The preceding theoretical analysis suggests that Erk activity shows a non-linear dependence on Erk expression levels, which makes it desirable to analyse cells with different Erk expression levels. In order to sort cells according to their expression level or to enable single cell analysis — which would allow the analysis of all sorts of expression levels — it is handy to express Erk together with a fluorescent label. Coexpression can be realised either by constructing a fusion protein of Erk and a fluorescent label or by expressing label and protein separately but with equal efficiency.

I have cloned a GFP-Erk2 fusion protein as well as a construct with separate expression of Erk and GFP, as there was some doubt as to whether a fusion protein would influence the mechanism of Erk activation by Mek. The separate expression with equal efficiency was realised by inserting the sequence of a so-called 2A peptide between the sequence of Erk and GFP. The peptide T2A originates from the insect virus *Thoseaasigna* and mediates the separate expression of contiguous peptides within the viral genome. While the ribosome translates the RNA sequence, it skips catalysis of the glycyl-prolyl peptide bond one amino acid before the C-terminus of a 2A peptide. Precisely spoken, cells with the EGFP-T2A-Erk2 construct express EGFP with a c-terminal T2A-tag and Erk2 with an additional proline at its N-terminus. Cleavage mediated by the T2A peptide (and other 2A peptides of different viral origin) is a universal phenomenon in eukaryotic cells and thus can be used to separately express two different genes with the same dosage. However, the mechanism does not have 100% efficiency and varies depending on the cellular system [81]. In Hek293 cells, 95% of the total amount of Erk2 is expressed separately (compare the band intensity of Erk2 with the intensity of the GFP-T2A-Erk2 band in Figure 3.10B).

I also cloned a plasmid with the EGFP-T2Arev-Erk2 sequence, where T2Arev stands for the reverse complement sequence of T2A. The reverse complement sequence contains a stop codon and thus the cells will only express EGFP with

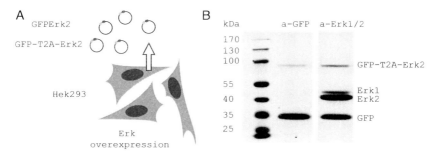

Figure 3.10   Expression of GFPErk2 and GFP-T2A-Erk2 in Hek293 cells. A) Hek293
cells were stably transfected with the fusion construct GFPErk2 or with the GFP-T2A-
Erk2 construct that mediates separate expression of GFP and Erk2 with equal efficiency.
B) Western blot analysis of Hek293 cells expressing GFP-T2A-Erk2. The blot was
probed against GFP (1st column), and probed with a pan-Erk antibody one day later
(2nd column = 2nd scan of first column). The two colour Li-COR system allows to see
that the GFP-T2A-Erk2 fusion protein is recognised by the pan-Erk and by the GFP
antibody (see also the Immunoblot protocol in section B.1).

a shortened tag. Cells expressing this construct have been used as a control
system. All lentiviral expression plasmids have been packed into HIV-based
replication-incompetent viruses. Hek293 cells were stably transfected by viral
transduction. More details on the cloning procedure, virus production and viral
transduction can be found in appendix section A.

Hek293 cells as experimental system

Hek293 cells had been immortalised by transformation with DNA that originates
from Adenovirus 5 [60]. The MAPK pathway can be studied at deactivation as
well as at activation after stimulation of Hek293 cells with growth factors [80].
From previous studies in our lab we know that the basal phosphorylation of Erk
is low, even in presence of FCS within the medium, which makes these cells ideal
candidates for analysis of Erk activity at weak pathway stimulation. Also, the
cells show a constitutively high level of AKT activation. Taken together this
suggests that their proliferative potential originates from modifications within
the PI3K/AKT pathway. When the cells drive proliferation via PI3K/AKT
signalling, an increased activation of Erk would not greatly affect the growth
properties of those cells. That means it is feasible to keep cells at different levels

of Erk expression over longer periods of time without selective pressure. This makes Hek293 cells an excellent test tube for understanding the correlation of Erk expression level and Erk activation - however, to understand physiologically relevant cell based decisions, it would be necessary to pursue the analysis in a different cellular system.

Proliferation is driven by PI3K signalling in Hek293 cells

At first I want to show that perturbing the expression level of Erk indeed does not influence proliferation of Hek293 cells. Therefore I have measured growth curves of WT Hek293 cells and of cells that overexpress either GFP, GFP-Erk2 or Erk2 and GFP separately. Growth was followed in real-time by using the impedance-based system xCELLigence [76]. In this assay the bottom of the culture dish contains gold electrode structures. Cells, which are basically insulators, increase the measured impedance of the circuit when they grow and spread on these electrodes. The technique is non-invasive and allows for keeping the cells at optimal growth conditions during the whole assay. However, the measured impedance is a complex property that also depends on the cell surface composition and the way cells interact with their substrate — it was observed, that a decrease or increase in impedance not necessarily reflects cell death or growth [8]. I have verified that during a proliferation assay the measured ci (= cell index, a background normalised measure of the impedance) represents the number of cells with sufficient accuracy, as outlined in the appendix section G.1. Figure 3.11A shows three replicates of growth curves for each cell line. For each growth curve I calculated the amount of doublings by i) normalising the cell index to a reference cell index at an early time point t_0 and ii) taking the logarithm to base two of this quotient:

$$\text{doublings} = \log_2 \frac{\text{ci}(t_{\text{end}})}{\text{ci}(t_0)} . \tag{3.8}$$

For each cell type the mean and the spread of the doubling events from three replicates is shown in Figure 3.11B. The slope of such a curve is proportional to the average growth rate, due to the following congruence:

$$\frac{1}{t_{\text{end}} - t_0} \int_{t_0}^{t_{\text{end}}} \frac{1}{\text{ci}(t)} \frac{\text{d}\,\text{ci}(t)}{\text{d}t} \text{d}t = \frac{1}{t_{\text{end}} - t_0} \ln \frac{\text{ci}(t_{\text{end}})}{\text{ci}(t_0)} . \tag{3.9}$$

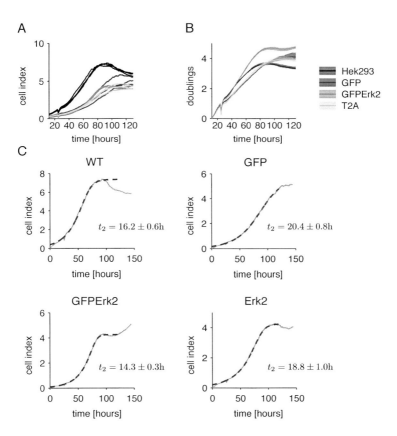

Figure 3.11 WT Hek293 cells or Hek293 cells expressing GFP, Erk2 or GFPErk2 have been seeded at $5 \cdot 10^3$ cells/well and growth was followed in real time using the impedance-based technology xCELLigence. A) Cell index raw data growth curves (3 technical replicates) and B) the cell index values transformed to the number of doublings according to equation (3.8). In B) the solid line corresponds to the mean of 3 replicates and the shade extends from the minimum to the maximum value. C) The Richards growth model (blue dashed lines) was fitted to the measured growth curves (grey lines). Shown is one exemplary fit (all fits in section G.2). The mean and standard deviation of the cycle time (calculated from the fitted model parameters according to equation (3.11)) is also shown.

By visual inspection it can be seen that the slopes are relatively alike for WT cells and cells expressing GFP or Erk2 and GFP separately. Only the cells with the Erk2-GFP fusion construct stick out with a higher growth rate. To leave an analysis that is only based on eye inspection, the results can be refined by fitting a growth model to the data. The cells grow exponentially, however at some point growth gets restricted due to the confined space in the culture well and progressing contact inhibition. This behaviour is registered by the empirical Richards model of growth [128], defined as follows:

$$\frac{d\,ci}{dt} = \frac{k \cdot ci}{1-m} \cdot \left[\left(\frac{G}{ci}\right)^{1-m} - 1\right] . \tag{3.10}$$

In this model the parameter G corresponds to the growth limit, the maximum ci that can be reached. The parameters k and m both influence the growth rate of the exponential phase. The duration of the exponential growth phase is determined by the shape parameter m. Actually, for $m = 2$ the model is equivalent to the logistic model, in which the growth rate ($d\,ci/dt/ci$) is linearly decreasing with the ci itself. For values of $m > 2$ the growth rate depends non-linearly on ci and the quasi-exponential growth phase is extended, followed by a steep decline of the growth rate towards values of ci close to the maximum G. Due to the empirical formulation of the model, a quantity like the characteristic cell cycle time cannot be derived from a single parameter of the model. However, it is possible to estimate a lower limit of the cell cycle time, for very small numbers of cells. For ci $\to$ 0 the model simplifies to $\frac{d\,ci}{dt} = \frac{k \cdot ci}{m-1}$ and so the minimal cell cycle time is

$$t_2 = \frac{m-1}{k} \cdot \ln 2 . \tag{3.11}$$

The analytical solution of the Richards model in eq. (3.10) reads

$$ci(t) = G\left[\left[\left(\frac{G}{ci_0}\right)^{m-1} - 1\right] \cdot e^{-kt} + 1\right]^{\frac{1}{1-m}} \tag{3.12}$$

with $ci_0 = ci(t = 0)$ and was used to fit the growth curves of the differently transfected Hek293 cells. All fits can be found in the appendix section G.2, Fig. 3.11C shows one exemplary fit for each cell line. The average minimal cell cycle time was estimated according to eq. (3.11) and is indicated at the top of each panel in Fig. 3.11C. Considering these numbers, still Hek293 cells

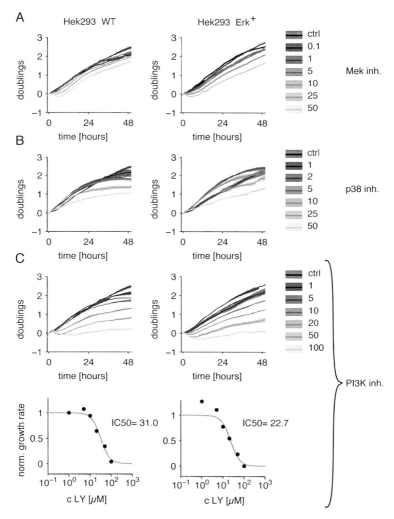

Figure 3.12 WT Hek293 cells (left column) or Hek293 cells overexpressing Erk2 (right column) have been seeded at $5 \cdot 10^3$ cells/well and treated with the indicated concentrations (in μM) of A) Mek inh. U0126, B) p38 inh. SB203580 or C) PI3K inh. LY294002 (top panels) or the solvent control DMSO 1 day after seeding. Growth was measured for two further days. The raw data have been transformed to doubling events since inh. application. The figures show the mean and the spread (min to max) of doubling events from 2 technical replicates. (Only one measurement of Mek inhibition in Erk overexpressing Hek293 cells.) In C) the average growth rates (normalised to the DMSO control) have been fitted according to eq. (3.13) to obtain the IC50 value.

stably transfected with the fusion protein GFPErk2 seem to grow fastest, with
an average doubling time of ≈ 14 hours, which is two hours less then the one of
regular Hek293 cells. However, cells expressing Erk2 and GFP separately grow
slower with an average cell cycle time of ≈ 19 hours, which leaves the analysis
inconclusive. Taking GFP expressing cells as a reference, it would seem that
the addition of GFPErk2, or of GFP and Erk2 separately, both confer a slight
growth advantage.

To gain further insight into the question, I took the reverse approach by in-
hibiting specific kinases to see whether it can compromise proliferation. Inhibition of p38 or Mek1/2 did not significantly affect proliferation of the cells (see
Fig. 3.12A and B), however inhibition of the PI3K was very efficient, causing
full growth arrest at a concentration of 100 μM (see Fig. 3.12C). The IC50 value
of the PI3K inhibitor was estimated from the average growth rates of the time
interval $t = [30, 54]$ (hours) at the various concentrations of the inhibitor by
regression with the Hill-slope model

$$\mathrm{Inh}_{\mathrm{rel}} = \frac{1}{1 + \left(\frac{c}{\mathrm{IC50}}\right)^n} \ . \tag{3.13}$$

The average growth rates were normalised with respect to the average growth
rate of the solvent control (DMSO) to obtain the relative inhibition $\mathrm{Inh}_{\mathrm{rel}}$ as
written in eq. (3.13). Actually, the literature is not clear about how to effi-
ciently and reliably estimate IC50 values of an inhibitor from growth curves. An
alternative approach is to use, e.g. the area under the growth curve to estimate
the effect of an inhibitor. I have analysed different methods of IC50 estimation
and found that the average growth rate is the most robust and reliable predic-
tor of the 4 different methods tested. This result and other general methods of
growth curve analysis shown in this chapter have been published recently [164].

The IC50 value of the PI3K inhibitor was found to be 31 μM in WT Hek293
cells and is comparable in cells overexpressing Erk2, with 23 μM. These results
suggest that indeed signalling via PI3K/AKT drives proliferation in Hek293
cells. Yet, an additional slightly positive influence of increased Erk levels cannot
be excluded, though this positive effect should have made Erk overexpressing
cells more robust to PI3K inhibition, which is obviously not the case.

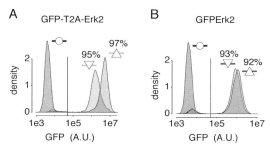

Figure 3.13 Stably transfected Hek293 cells had been sorted into fractions of low and high GFP label intensity by FACS and cultured thereafter. Shown is the distribution of GFP label intensity 9 days after sorting in WT Hek293 cells (pink) and in Hek293 cells expressing low (green) or high (blue) levels of A) Erk2 and B) GFPErk2. The vertical line is the gate separating GFP$^-$ and GFP$^+$ cells. The fraction of GFP$^+$ cells is indicated with the percentage next to the densities. Cells according to this distribution have been subjected to an Immunoblot analysis as shown in Fig. 3.14A. The coloured triangle and circle symbols label the different distributions here and in Fig. 3.14.

3.4 The phosphoErk response is stronger but not prolonged in Erk overexpressing cells

The phosphoErk response to EGF stimulation

To enable the analysis of cells at different levels of Erk2, cells expressing either GFPErk2 or Erk2 (T2A construct) have been sorted into two groups of different degrees of Erk overexpression according to the intensity of the GFP label using FACS. The sorted cell fractions ($\approx 1 \cdot 10^5$ cells per fraction) were cultured and one week thereafter enough cell material was available for experiments.

The Hek293 cells have been kept in medium supplied with 10% of FCS (fetal calf serum). As FCS contains several growth factors, cells are usually serum-starved before targeted stimulation of the MAPK pathway by EGF. However, the Hek293 cells contain no detectable levels of phosphoErk when they are cultured in the presence of 10% FCS. Also, cells that overexpress Erk2 have stationary phosphoErk levels below the detection limit (no phosphoErk1/2 bands at t=0 in Western blots in Fig. 3.14A and 3.15A). For that reason the cells were not serum-starved before EGF treatment. To compare the response of Erk phosphorylation in WT and Erk2 overexpressing Hek293 cells, the cells were treated with 25 ng/ml EGF and lysed at various time points thereafter, covering 1 min to 30 min after stimulation. The lysate was subjected to a Western blot analysis of total Erk1/2 and dual phosphorylated Erk1/2 (see Fig. 3.14). Also, on the same day of the experiment, cells of each type were left untreated and harvested for live single cell analysis of the GFP label on a flow cytometer (shown in Fig. 3.13). More than 90% of the sorted cells are GFP positive. Nine days after sorting, the distribution of the GFP label in the low and high Erk2 expressing group overlaps, however, the median of GFP intensity of the groups is distinct.

The average degree of Erk overexpression has been quantified from the Western blot (see Fig. 3.14B) and was 7x (9x) in the "low Erk2" and 15x (21x) in the "high Erk2" group of cells that overexpress GFPErk2 (Erk2 from T2A construct). The phosphoErk2 response over time is shown in Fig. 3.14C. All cell lines have their maximal response 5 to 7.5 min after stimulation. The amount of phosphoErk2 at the peak increases with Erk2 overexpression, however, the exact foldchange of Erk2 phosphorylation when comparing between WT and overexpressing cells cannot be estimated from this analysis, as each cell line was analysed on a separate Western blot. What can be compared is the level of phosphoErk2 normalised to maximal phosphoErk2, as shown in Fig. 3.14D. In this presentation it is revealed, that the kinetics of Erk2 phosphorylation is surprisingly similar in all overexpressing cell lines. Whether the cells express the fusion protein GFPErk2 or Erk2 alone is irrelevant.

The WT Hek293 cells in contrast stand out, as relative Erk2 phosphorylation stays elevated for a much longer time. In general, it takes roughly one hour until phosphoErk levels go back to untreated levels (see additional data in section B.2).

The experiment was repeated, this time in a manner that allows the estimation of the foldchange in Erk phosphorylation at overexpression. The lysates of EGF stimulated WT Hek293 cells (2 biol. replicates) and of Erk2 overexpressing Hek293 cells of the "high Erk2" group (2 biol. replicates) have been analysed on separate blots, however this time one and the same control lysate was loaded on each of these blots (framed in red in Fig. 3.15A) as a means for normalisation of the blots with respect to each other. Erk2 was overexpressed between 19 and 20x on average (Fig. 3.15B). Total phosphoErk in WT Hek293 cells was quantified with ≈ 5 A.U. at the peak (Fig. 3.15C), whereas phosphoErk maximally reached values around 120 A.U. in the overexpressing cells (Fig. 3.15D). It means that at the time of maximal Erk activation, phosphorylation is increased roughly 24x, which is in the order of magnitude of Erk overexpression. This

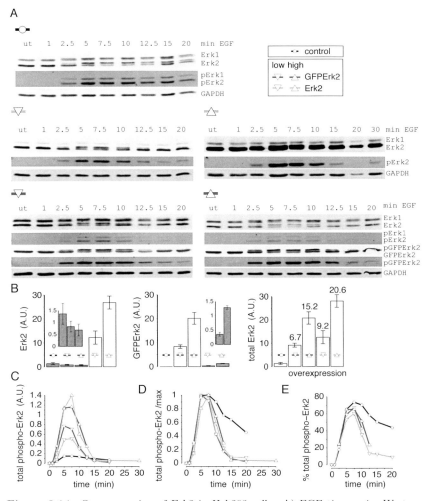

Figure 3.14 Overexpression of Erk2 in Hek293 cells. A) EGF time series Western blot analysis of Erk1/2, phosphoErk1/2 and GAPDH in Hek293 cells with and without overexpression of Erk2 as indicated by the legend. B) Quantification of Erk overexpression by evaluation of band intensities of blots shown in A). The bands of Erk2 and GFP(T2A)Erk2 were normalised to the intensity of the Erk1 band. To obtain the total amount of Erk2, normalised intensities of Erk2 and GFP(T2A)Erk2 band were added up. C) Quantification of total amount of phosphoErk2 (normalised by GAPDH) from the 4 different Western blots in A). D) The same as in C), but normalised by the maximal value of phosphorylation. E) Intensity of phosphoErk2 and Erk2 as quantified by the pan-Erk antibody is used to determine relative Erk2 phosphorylation (in %).

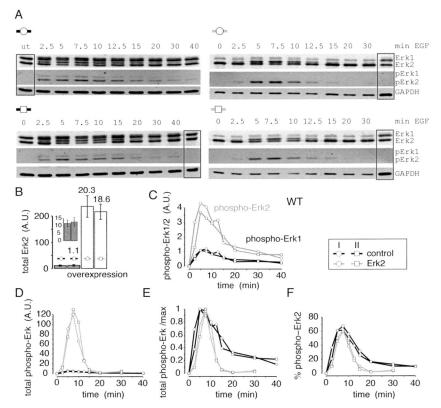

Figure 3.15 EGF time series Western blot analysis of Erk1/2, phosphoErk1/2 and GAPDH in Hek293 cells without transfection (2 biological replicates) and with over-expression of Erk2 (2 biol. replicates), as indicated by the legend. To compare the results of different Western blots the lysate of untreated Hek293 cells (replicate I) was analysed on each blot (indicated with the red frame) and used for normalisation in the densitometric analysis in the following. B) The fold change in Erk2 expression is quantified from the band intensity of Erk2 plus band intensity of GFPT2AErk2. C) The phosphoErk1 and phosphoErk2 kinetics after EGF stimulation in untransfected Hek293 cells (2 biol. replicates as quantified from A)). D) Total Erk phosphorylation in normal and Erk2 overexpressing Hek293 cells. E) The same as D), but normalised to maximal phosphorylation. F) Relative phosphorylation of Erk2 as quantified from the Erk2 and phosphoErk2 band detected by the pan-Erk antibody.

means that at maximal activation, the amount of phosphoErk increases linearly with the amount of available Erk. However, already 20 min after stimulation the levels of phosphoErk are almost identical (Fig. 3.15D). Obviously, Erk is dephosphorylated much faster in Erk overexpressing cells. This difference in Erk deactivation kinetics can be seen very well, when the time course is normalised to the value of maximal phosphorylation, as shown in Fig. 3.15E.

As the phosphorylated Erk1/2 band is shifted in gel electrophoresis, the phosphorylated and dephosphorylated band can be quantified with the Erk1/2 antibody separately, if the bands are sufficiently resolved. This allows an estimation of % Erk2 phosphorylation, as shown in Fig. 3.15F. With this quantification it can be verified, that % Erk2 phosphorylation is similar in WT and Erk2 overexpressing cells at the peak, but reduced much faster thereafter in Erk2 overexpressing cells. Mek seems to be far from saturation, as it can catalyse the activation of a 20x increased amount of Erk, however in this experiment, the peak of Erk activity is slightly delayed in overexpressing cells (Fig. 3.15E). Given the situation, several questions arise. What is the reason for the increased speed of Erk deactivation in overexpressing cells? Is it the action of a negative feedback and if so which one is responsible? Can kinetic effects of Erk activation by Mek play a role? At first I investigated the phosphoErk response on the single cell level to find out more about the potential non-linear correlation of Erk expression and activation.

The phosphoErk response at the single cell level

Single Hek293 cells were analysed by flow cytometry after fixation, permeabilisation and immunostaining with antibodies targeting total Erk1/2 and phosphpo-Erk1/2. The single cell data were preprocessed to exclude cellular debris as well as cell clumps from the analysis. If necessary, fluorescence compensation was performed to remove artefacts resulting from fluorescence spillover (the principle of fluorescence compensation is described in appendix section E.2). As cells vary in size and granularity, all antibody signals are expected to be correlated to a certain extent, i.e., bigger cells will also have a higher protein content. To correct antibody signals for the influence of cell morphology I have transformed the data according to an approach that is based on the 3D regression of fluorescence signals (FL parameters) over the scattered light parameters FSC and SSC, using the MATLAB functions published and described by Knijnenburg et al. [86].

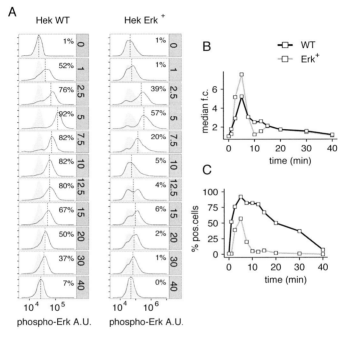

Figure 3.16   Erk phosphorylation measured by flow cytometry. A) Distribution of the phosphoErk-Alexa647 signal (FL4.A) in EGF treated Hek cells (WT or Erk[+], grey boxes indicate time after stimulation, green dashed line indicates the median). Percentages in the top right corner refer to the % of phosphoErk-positive cells, when all unstimulated cells below the 99% quantile are arbitrarily set to phosphoErk-negative. The populations of Erk overexpressing cells contain only transfected cells (gate for GFP positive). B) Fold change of the phosphoErk-signal median as indicated by the dashed line in panel A) as a time series. C) % positive cells as indicated with the percentages in panel A) as time series.

This data transformation has at least two advantages. Firstly, the correlation of two measured fluorescence parameters can be estimated more realistically and secondly, the signal of one fluorescence parameter will show the variability caused by gene expression noise and transfection efficiency rather than the variability caused by cell morphology. Preprocessing of the flow cytometry data is described in more detail in the appendix section E.1.

Figure 3.16A shows the distribution of phosphoErk in single cells after stimulation with EGF in a time series. The median of each distribution is indicated with a dashed green line in panel A) and panel B) summarises the fold change of the median (with respect to t=0) in Hek293 WT and Erk2$^+$ cells. The data confirm results of the Western blot analysis — WT and Erk2$^+$ cells show their maximal response ≈5 min after stimulation, however the transition from the active state back to unstimulated niveau is much sharper in Erk2 overexpressing cells. The data show slight bimodality at early time points after stimulation, which indicates a fraction of non-responding cells. Interestingly, bimodality persists also for late time points in Erk2 overexpressing cells. One of the populations is aligned with the unstimulated distribution, whereas the second population is shifted to higher phosphoErk values. This might indicate two populations of cells with different kinetics of Erk deactivation or a damped oscillation in Erk deactivation. If this behaviour is linked to the Erk expression level, will be seen later.

Erk overexpressing cells show a greater variablilty of their phospoErk signal with a distribution ranging over ≈ 1.5 orders of magnitude, compared to WT Hek293 cells with a variability of only one order of magnitude (see first panel for t=0 in Fig. 3.16A). It is not possible to determine which cells are actually phosphoErk positive and which are negative. In order to characterise the fraction of cells with a significant shift of phosphoErk, I have arbitrarily decided to set cells below the 99% percentile of the distribution at t=0 to "phosphoErk-negative cells". The so calculated fraction of "phosphoErk-positive cells" is indicated in the top right corner of each panel in Fig.3.16A and shown as a time series in 3.16C. The much faster downregulation of phosphoErk in Erk2 overexpressing cells is even more apparent in this parameter. At t=10 min the phosphoErk distribution can hardly be distinguished from the state at t=0 in Erk overexpressing cells, which results in an almost identical median and fraction of positive cells (when comparing t=0 and t=10mins). Due to bimodality, the median is slightly increasing for timepoints thereafter (3.16B)— however, the fraction of "positive cells" stays low (3.16C).

All in all the results of this analysis are in line with the Western blot results. How do the phosphoErk distributions now correlate with the expression level of Erk? In WT Hek293 cells there is always a slight correlation of Erk and phosphoErk signals (see Fig. 3.17). The unstimulated state at t=0 shows the minimal Pearson correlation of 0.44. With increasing Erk activity the correlation coefficient increases up to 0.76 at 7.5 min after EGF stimulation.

The Erk2 overexpressing cells analysed here have not been sorted for a certain expression level. They are a very heterogeneous population that also contains a small share of untransfected cells. I show the correlation of Erk expression level and Erk phosphorylation for transfected (GFP$^+$) and untransfected (GFP$^-$) cells separately, in Fig. 3.18A and B, respectively. Calculation of a correlation coefficient for the untransfected cells was not done, as the cell count is relatively low. However, as the transfected and untransfected cells are derived from one and the same sample, measured fluorescence intensities are directly comparable between WT and Erk2$^+$ cells. While the fold change in Erk expression between two cells can be as much as 30x in WT cells, it reaches a factor of 100 in Erk over-expressing cells. The measured phosphoErk levels range from 10^4 to 10^5 A.U. in unstimulated WT cells, while they can go up to $10^{5.5}$ A.U. in Erk2$^+$ cells. This means that even at weak pathway stimulation phosphoErk is elevated in Erk overexpressing cells. However, as the rather small correlation coefficient of 0.27 indicates, there is partial robustness of phosphoErk with respect to the amount of Erk. Just like in WT cells, the correlation of phosphoErk with Erk is maximal during maximal pathway stimulation, the correlation coefficient reaches 0.49 5 min after application of EGF. Interestingly, the population of non-responding cells at 5 and 7.5 min after stimulation shows equally low phosphoErk levels independent of Erk expression level. Also for later time points partial robustness of phosphoErk becomes visible. The bimodality in phosphoErk, which is especially eminent at 12.5 and 15 min after treatment with EGF (see Erk$^+$ cells in Fig. 3.16A) can also be seen in the scatter plots of Erk and phosphoErk. The two populations do not distinguish themselves by the Erk expression level, which means that their different phosphoErk response is not caused by some kinetic mechanism directly depending on Erk expression level. It remains an open question what causes these two populations to appear. I have used an expectation maximisation algorithm of Gaussian mixture modelling to cluster the data according to 2 Gaussian distributions, as shown Fig. 3.18C. At 15 min

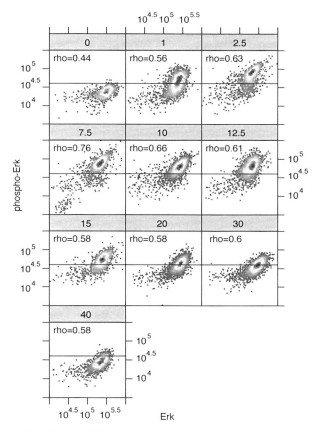

Figure 3.17 Correlation of Erk expression and Erk phosphorylation in WT Hek cells after stimulation with EGF. Flow cytometry was used to measure the Erk and phosphoErk signal of single Hek293 cells. The number at the top of each panel indicates the time [min] after stimulation with EGF. Intensity of Erk and phosphoErk signals in A.U. on a logarithmic scale. The red line marks the 99% percentile of the phosphoErk signal in untreated cells. Events below the 99% percentile of the background signal in each parameter are not shown. rho indicates the calculated correlation coefficient of the distributions.

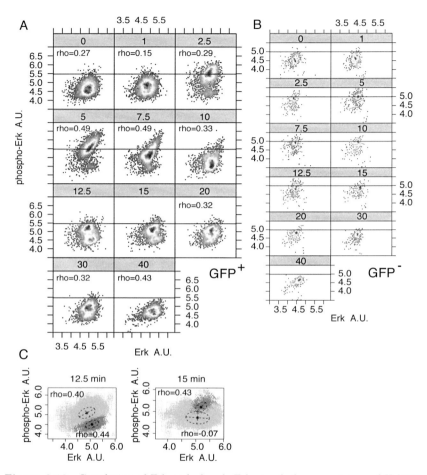

Figure 3.18 Correlation of Erk and phosphoErk signals from a mixture of Hek293 cells with and without transfection with GFP-T2A-Erk2. The cells were separated for Erk$^+$ cells (GFP$^+$ cells) in A) and WT cells (GFP$^-$) cells in B) *in silico*. The number at the top of each panel indicates the time [min] after stimulation with EGF. Intensity of Erk and phosphoErk signals in A.U. on a logarithmic scale. The red line marks the 99% percentile of the phosphoErk signal in untreated cells. Events below the 99% percentile of the bg signal in each parameter are not shown. C) The 2D distribution of Erk and phosphoErk signals at 12.5 and 15 min after stimulation in Erk overexpressing cells (as shown in A) was clustered into two 2D Gaussian distributions. The semi-major and semi-minor axis of the black ellipse correspond to the eigenvectors of the covariance matrix of the respective distribution. rho indicates the calculated correlation coefficient of the distributions in A) and C).

there is full robustness of Erk activity in the population of lower phosphoErk levels, as indicated by the correlation coefficient of -0.07.

In summary, the single cell data confirm that at low Erk activity the phosphoErk level is partially robust to the Erk expression level. Yet, we still don't know how Erk activity is downregulated so much faster in Erk2$^+$ cells. The investigation follows in the next chapter.

3.5 Sequestration effects and fast feedbacks can explain the transient phosphoErk response

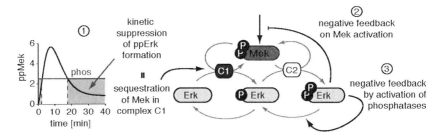

Figure 3.19 Mechanisms that can contribute to faster decline of Erk activity at Erk overexpression. ① Kinetic suppression of ppErk formation can result from sequestration of Mek in complex C_1 as soon as the deactivating phosphatases have a higher catalytic activity than the activating kinase Mek. The time dependent activity profile of the kinase (black) and the phosphatase (grey) after stimulation with EGF are schematically shown in the panel on the left. The area shaded in grey indicates the times when the condition for mechanism ① is fulfilled. Additionally an Erk dependent negative feedback can reduce the activity of Mek (mechanism ②). It is known that Erk also regulates the activity of phosphatases (mechanism ③).

Three very simple mechanisms, which could explain faster deactivation of Erk in cells with higher Erk expression level, come to mind. I have shown previously that the sequestration of active Mek in complexes with unphosphorylated Erk can restrict the formation of dual phosphorylated Erk as soon as the maximal turnover rate of the phosphatase(s) exceeds the maximal turnover rate of the kinase Mek. I describe this mechanism as "kinetic suppression of ppErk formation", which is illustrated as the 1st mechanism in Fig. 3.19. Assuming that the activity of phosphatases is constant during the first 40 minutes after

EGF stimulation (grey line), but the activity of Mek (black line) shows a time dependent profile, the condition $v_{\text{max,Phos}} > v_{\text{max,Mek}}$ will be fulfilled for very early time points and again for later time points after stimulation (area shaded in grey). The relatively equal amounts of dual phosphorylated Erk in WT and Erk2$^+$ cells already 20 min after stimulation might be explained by the Erk concentration dependent suppression of Erk activation as soon as the activity of Mek has sufficiently declined. The most simple model that can reproduce this effect appreciates the enzyme-substrate complexes of (p)Erk with Mek, C_1 and C_2, but assumes that phosphatases in contrast never get saturated (see pathway scheme in Fig. 3.19). This might be a valid hypothesis, because apart from the DUSPs (or MKPs), whose expression is controlled by Erk activity, several other phosphatases contribute to Erk inactivation. In particular PP2A, a negative regulator of Mek as well as of Erk [108] is ubiquitously expressed at high levels [12].

The activation of phosphatases in an Erk dependent manner could also explain faster downregulation of Erk activity in Erk overexpressing cells (mechanism 3 in Fig. 3.19). Erk induces the expression of DUSPs and may also prevent degradation of DUSPs. However, both of these processes require the synthesis of new proteins and thus would be to slow to explain an effect already 20 min after stimulation. Possibly, some phosphatases are activated by a post-translational mechanism in an Erk dependent manner. Alternatively, it is possible that Erk overexpressing cells show elevated basal phosphatase activity compared to WT cells, as they have perceived higher levels of Erk activity previously.

Apart from negative feedback regulation via phosphatases, there is also a fast post-translational negative feedback that acts on Raf and leads to reduced levels of ppMek (mechanism 2 in Fig. 3.19).

The first of the three hypotheses requires that phosphorylation of Erk by Mek follows a distributive scheme. As outlined in the introduction, most experimental studies hint at this scheme, however there is still doubt on the *in vivo* relevance of this phenomenon. I tried to investigate the activation mechanism of Erk as described in the next section.

Is Erk phosphorylation by Mek distributive?

The easiest way to show whether phosphorylation of Erk is distributive or processive in mammalian cells would be to measure single phosphorylated Erk as well as dual phosphorylated Erk. The distributive model predicts an accumulation of pErk before significant amounts of ppErk can be formed. Also, when Erk is overexpressed, pErk should accumulate at weak levels of active Mek.

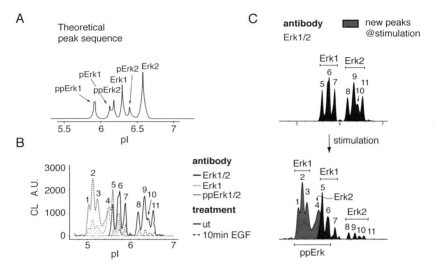

Figure 3.20   PhosphoErk1/2 analysis of Hek293 cells using isoelectric focusing. A) The theoretical isoelectric point (pI) pattern of the 6 different phosphoErk forms as supplied by proteinsimple™. The sketch was redrawn from the illustration found on their web site. B) The protein lysate was resolved by isoelectric focusing and probed with antibodies specific for Erk1/2, Erk1 and ppErk1/2. The different antibodies are indicated by different colours. Spectra are shown either with solid lines (untreated Hek293 cells) or with dashed lines (Hek293 cells stimulated with EGF for 10min). C) Only the spectra obtained from the Erk1/2 antibody, upper panel unstimulated cells, lower panel treated cells. Novel phospho-forms that appear at stimulation have been coloured in red.

How to detect single phosphorylated Erk? The most sensitive approach would be mass spectrometry, however, we did not dispose of this option.

Commercial antibodies for the detection of either pThr- or pTyrErk are available, however, despite their high sensitivity, these antibodies have been shown to show a high degree of "cross-specificity" for unphosphorylated and dual phosphorylated Erk (false positives), which limits their use in a standard Western blot approach [124]. Even if the antibodies had a 100% specificity, they still could not be used to quantify the relative amounts of the differently phosphorylated states, as every antibody has its unique affinity to a target protein. That means, e.g., the band intensity from an anti-pThrErk antibody and an anti-pThrpTyrErk antibody cannot be compared. Thus it is necessary to i) separate the differentially phosphorylated Erk isoforms by some means (mass or charge), ii) use the phospho/isoform-specific antibodies to identify the different forms and to iii) quantify the relative amounts of the forms with a pan-Erk antibody with specificity to a site outside of the activation loop.

Already the separation step is a challenge. On a regular Western blot, phosphorylated Erk is shifted to higher protein mass in location compared to unmodified Erk. However, a differentiation of the various phospho-forms is not feasible. A relatively novel method has been introduced in 2006, where the polyacrylamide gel is supplemented with so-called Mn^{2+}-Phos-tag molecules, which reversibly capture phosphomonoester dianions and thus slow down the movement of phosphorylated proteins compared to nonphosphorylated species during SDS-PAGE [82]. We tried to establish the method in our lab, however we were not able to obtain reproducible band patterns.

For that reason I decided to switch to an assay with charge-based separation. Proteins were separated by their isoelectric point using a capillary-based technique [116]. The method is described in more detail in appendix section D. The analysis of Erk phosphorylation has been performed previously and the manufacturer of the assay has documented spectra of the form as shown in Fig. 3.20A. The spectra turned out to be more complex in this analysis. In Hek293 cells which were kept in medium containing 10% FCS, Erk1 and Erk2 each can be found in 4 different forms (see Fig. 3.20B and C, peak 6 seems to consist of two tightly spaced peaks). Whether one of these forms represents the single phosphorylated form of Erk1/2, respectively, cannot be said. The antibody for ppErk1/2 does not recognise a target substrate (see Fig. 3.20B) and despite the likely cross-specificity of the antibody for single phosphorylated Erk, its sensitivity for this substrate might not be sufficient. At stimulation of the pathway 4 new peaks show up in the spectrum (Fig. 3.20B and C). All of these 4 peaks are recognised by the antibody which is specific for ppErk1/2. Peaks one to three clearly represent three different forms of ppErk1 (and of pErk1 when the level of pErk1 was below the detection range of the ppErk1/2 antibody without stimulation). Peak number 4 seems to represent a form of ppErk2 or pErk2. Peaks 5 and 6 are special in that they are clearly a mixture of different species. Peak 5 is only recognised by the Erk1 antibody in unstimulated cells and at stimulation the Erk1 signal becomes weaker (see Fig. 3.20B). This would mean that this peak constitutes a form of pErk1 (not recognised by the ppErk1/2 antibody) or of Erk1 and shifts to smaller pI values due to phosphorylation. However, at the same time, this peak gains in size at stimulation and is recognised by the ppErk1/2 antibody. This means that some form of ppErk2 superimposes the pErk1/Erk1 peak. The same seems to be true for the (double-)peak 6, however here the overall signal intensity is decreasing at stimulation. Peak number 7 seems to be one form of Erk1 which is decreasing at stimulation.

As long as Erks only modification lies within the activation loop, one can expect 6 distinct peaks and their order and position is straightforward, because every single phospho-group contributes to a shift to a lower charge of defined dimension (see scheme 3.20A). The 6 distinct pI values allow a full resolution of all Erk species. In Hek293 cells, obviously other modifications of Erk are present. The first consequence is that peak identification would require extended experiments in which the cell lysate is treated with agents that specifically remove certain types of modifying groups. Secondly, only when the other modifications are non-phospho groups, one could pretreat the samples to eliminate all overlapping/unresolved peaks for full quantification.

In conclusion, it would take a longer period of experimentation to successfully use isoelectric focusing for resolving all different Erk1/2 phospho-forms which is not within the scope of this thesis.

Is there a negative feedback from Erk to Mek in Hek293 cells?

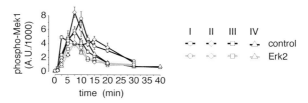

Figure 3.21 Relative phosphorylation of Mek1 was measured in 4 biological replicates of Hek293 WT (control) and Erk2$^+$ cells with the bioplex assay. Measurement III and IV were obtained from the same proteins lysates as the phosphoErk measurement shown in Fig. 3.15. Error bars indicate the 95% confidence interval (see also appendix section C.2).

When Erk was knocked down in several different colon carcinoma cell lines, phosphorylation of the remaining Erk was upregulated due to a negative feedback that acts on the level of Raf. As Erk exerts this feedback by phosphorylation of specific sites in Raf, this mechanism is fast enough to be relevant within the first minutes after EGF stimulation. I have measured dual phosphorylation of Mek1 in response to EGF in WT and Erk2$^+$ cells (Fig. 3.21).

Relative phosphorylation of Mek1 various times after stimulation with EGF is quite variable in the 4 biol. replicates of WT Hek293 cells, however relative phosphorylation of Mek1 is clearly reduced in Erk$^+$ cells (see Fig. 3.21). This reduction of Mek1 phosphorylation is significant and apparent already 2.5 min after stimulation. The straightforward explanation is the action of a very fast (post-translational), not necessarily direct, negative feedback from Erk to Mek. Whether the feedback regulation of Mek activity is sufficient to explain the faster deactivation of Erk in Erk$^+$ cells, will be analysed with the help of mathematical modelling in the next section.

The contribution of the mechanisms analysed by mathematical modelling

I have used mathematical modelling to test whether the different mechanisms suggested above can explain the observed kinetics of Erk phosphorylation. In this model fitting approach, I have used the measured ppMek1 signal as the input that leads to activation of Erk. For that, the analytical function

$$f(t) = a \cdot \left[e^{-\lambda t} \cdot (t^n - c) + c \right] \tag{3.14}$$

was fitted to the ppMek1 data. The respective fits are shown in the top panel of Figure 3.22. The different models are evaluated based on their ability to reproduce the according kinetics of Erk phosphorylation, as indicated with the question marks in the lower panel of Fig. 3.22.

When using the fitted ppMek data as input, the feedback to Mek is implicitly contained within the model. In order to test the mechanism of kinetic suppression of ppErk formation (mechanism ①) alone, the ppMek input function of WT Hek293 cells was used to fit the ppErk data of WT and Erk2 overexpressing cells. The model was formulated as depicted in Fig. 3.23A, with a dephosphorylation rate proportional to the amount of (p)pErk itself and a phosphorylation rate that

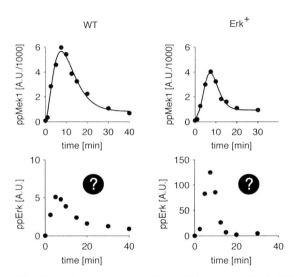

Figure 3.22 Which mechanism explains the ppErk time-profile in WT and Erk2[+] cells? The Bioplex ppMek1 data have been fitted using the analytical function (3.14). Parameters for WT: λ=0.28, n=2.05, c=1.26, a=650.68 and in Erk2[+] cells: λ=0.74, n=5.34, c=57.55, a=16.22. Fitting of mechanistic differential equation models is used to elucidate the experimentally observed phosphoErk kinetics (panels at the bottom), when the analytical fit of the ppMek1 data is used as input.

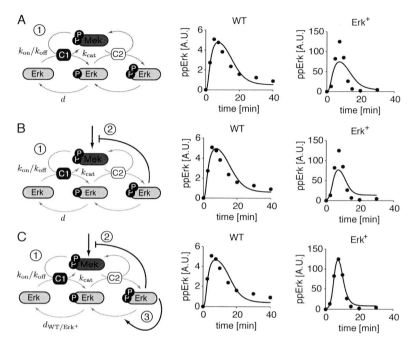

Figure 3.23 Model selection. A model of Erk activation has been expanded in 3 steps to test which model (solid lines in mid and right panels) can explain the experimental data (black dots in mid and right panels). All models include distributive dual phosphorylation by Mek where the enzyme substrate complexes C_1 and C_2 are appreciated (no Michaelis-Menten approximation). Identical rate constants for the first and the second phosphorylation cycle were assumed. At high expression levels of Erk sequestration of ppMek in complex C_1 can lead to suppression of Erk activation (mechanism ①). The model in A) has 4 parameters and ppMek1(t) measured in Hek293 WT cells was used as the time dependent stimulus. In model B) the negative feedback on Mek activation was added (mechanism ②) by using the cell type specific ppMek1 time profiles (shown in Fig. 3.22) as input. No new parameters are required. Model C) additionally assumes that dephosphorylation rates differ in WT and Erk$^+$ cells (mechanism ③) which extends the model to depend on 5 parameters. Model equations in appendix section I.4.

is based on the complex formation of ppMek and (p)Erk. It was assumed that the first and second (de)phosphorylation step are characterised by the same rate constants, so that altogether 4 parameters have to be fitted, the dephosphorylation rate d, the rate of catalysis of ppMek, k_{cat}, and the rates which describe the reversible complex formation, k_{on} and k_{off}. The model equations can be found in section I.4 and the method for parameter estimation is described in section I.5. The model is capable of a qualitative description of the ppErk kinetics in WT cells, but fails to explain the high ppErk peak in Erk2 overexpressing cells (see Fig. 3.23A).

Next, the model was extended to include the feedback on ppMek (mechanism ②, Fig. 3.23B), simply by using the cell type specific ppMek1(t) input functions. Slight changes in the shape of the ppErk fit are visible (Fig. 3.23B), however, also this model fails to explain the experimental data. Lastly, the model was further extended to also account for regulation of the phosphatase activity (mechanism ③). For simplicity it was assumed that the dephosphorylation rates are different in WT and Erk2$^+$ cells, which extends the model to a 5 parameter model — instead of d, I fit rates d_{WT} and d_{Erk+}. This model is capable of explaining the Erk phosphorylation kinetics (Fig. 3.23C).

Intuitively one could think that the dephosphorylation rate should be greater in Erk2$^+$ cells. However, the fit shows that the relation is reciprocal: a successful model fit is characterised by smaller dephosphorylation rate constants in Erk2$^+$ cells (see Fig. 3.24A, the two dephosphorylation rate constants are strongly correlated, read more about non-identifiability in appendix section I.7). A closer look at the sequestration of ppMek by Erk reveals why this is the case. Around 20 min after stimulation and later the Mek activity declines and significant amounts of phosphoMek become sequestered in the complex C_1 (see Fig. 3.24B), which suppresses the formation of ppErk and thus explains low levels of ppErk even though the dephosphorylation rate is lower in Erk$^+$ cells than in WT cells around that time (see Fig. 3.24C). In contrast, at the Erk activity peak, ppMek can work unobstructed and the maximal amount of formed ppErk is limited by the activity of the phosphatase. To achieve a phosphorylation peak as high as seen in the Erk2$^+$ cells data, the dephosphorylation rate must be relatively low. To further confirm that the sequestration of ppMek in C_1 is key to the ability of this model to fit the data, I have tested a simpler model, which also incorporates the feedback to Mek and allows different cell type dependent dephosphorylation rates, but in which phosphorylation is just linearly proportional to the amount of ppMek and (p)Erk (see appendix section I.6). This model cannot explain the data.

Though the smaller dephosphorylation rate in overexpressing cells makes sense in the model described above, it is unlikely that the $Erk2^+$ cells have lower basal phosphatase activity levels. What could be the reason for lower dephosphorylation rates around the time of maximal pathway activity? One explanation is that it is the other way around: dephosphorylation rate is "normal" at the peak of Erk activity, and then later upregulated in an Erk dependent manner by some fast post-translational mechanism. The other explanation is that the phosphatase activity is lower than expected at the peak of Erk activity, because also the phosphatase can get saturated.

To test this hypothesis, I formulated a model in which the dephosphorylation rate is modelled by Michaelis-Menten kinetics. Instead of parameters d_{WT} and d_{Erk^+}, the model contains the K_M value and the maximum turnover rate of the phosphatase, v_{max} (see Fig. 3.25A). This model, though having the same number of parameters, yields a qualitatively better fit of the $Erk2^+$ cells ppErk data, especially for the later time points (compare fits in Fig. 3.23C with fits in Fig. 3.25B and C). The K_M value is identified with a concentration around 27 A.U. , which indicates that at Erk overexpression the dephosphorylation rate does no

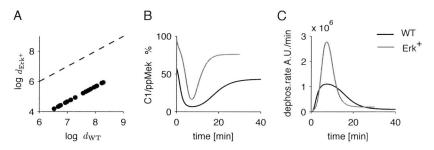

Figure 3.24 Analysis of the model that explains the data. Various features of the model C in Fig. 3.23, which can explain the experimental data, are analysed here. Panel A) shows the correlation of identified dephosphorylation rate constants in WT and Erk^+ cells from the best 20 fits (black dots). The dashed line shows the function $d_{Erk^+} = d_{WT}$. Panel B) shows the time dependent percentage of ppMek which is sequestered in complex C_1 for WT cells (black) and Erk^+ cells (grey) for the best parameter fit. Parameters of the best fit were used to draw the time dependent dephosphorylation rate $d \cdot ppErk(t)$ for both cell types in panel C).

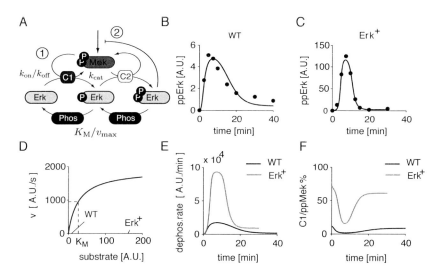

Figure 3.25 A model where the phosphatase can saturate is superior. In this model of distributive Erk activation, deactivation was assumed to proceed with Michaelis-Menten kinetics. The dephosphorylation reaction is characterised by K_M and v_{max}, as shown in the model scheme in panel A). The best model fit (solid line) of the WT and Erk$^+$ data (black dots) is shown in panel B) and C). Panel D) illustrates the catalytic activity of the phosphatase as a function of the substrate concentration with K_M and v_{max} as identified by the best model fit. The labels indicate the concentration of Erk in WT and Erk$^+$ cells. The predicted dephosphorylation rate is shown in panel E), sequestration of Mek in complex C_1 after stimulation with EGF in panel F). Model equations in appendix section I.4.

longer increase proportional to the amount of Erk (Fig. 3.25D). In numbers, the dephosphorylation rate is only increased by $\approx$ 5.4x at the ppErk peak, whereas the Erk level is increased about 20x. However, the dephosphorylation rate in absolute terms is higher in Erk2$^+$ cells than in WT cells the whole time in this model, which is in agreement with the intuition (Fig. 3.25E). While this model assumes that not only the kinase, but also the phosphatase might saturate with its substrate, it was ignored that the reason for the phosphatase saturation might be the high expression levels of Erk itself. That's why I finally tested a model, in which both the phosphatase and the kinase are modelled as molecules that may form complexes with the differently phosphorylated forms of Erk (Fig. 3.26A). Here the sum of free ppErk and ppErk bound to the phosphatase (complex D_2) has been trained to the experimental ppErk data.

Though this model shows improvement with respect to its ability to explain the WT data (Fig. 3.26B), I was not interested in showing an improved model, but rather to reveal how exactly this model explains the data. This model corresponds to the one, which was discussed at length at the beginning of this chapter — here either the formation of ppErk is limited due to sequestration of the kinase (C_1 = ppMek-Erk accumulates), or the formation of Erk is limited due to sequestration of the phosphatase (D_2 = Phos-ppErk accumulates), depending on the activity ratio of kinase and phosphatase. The predicted activity ratio of kinase and phosphatase after stimulation with EGF is shown in Fig. 3.26D for WT and in Fig. 3.26E for Erk$^+$ cells.

While at normal Erk expression levels neither the kinase nor the phosphatase gets anywhere near saturation (black lines in Fig. 3.26F and G), sequestration of ppMek in complex C_1 or of the phosphatase in D_2 switches depending on kinase to phosphatase activity ratio in Erk2$^+$ cells (grey lines in the same panels). This i) explains the high Erk phosphorylation levels at the peak and ii) the very low Erk phosphorylation levels already 20 min after stimulation. Also with this model, dephosphorylation rates are higher in absolute terms throughout the whole time in Erk2$^+$ cells. Interestingly, while the model explains the ppErk levels $\approx$ 5 min after stimulation it also predicts that a great amount of this ppErk will not be available for substrates, as it is sequestered in complexes with the phosphatase (see Fig. 3.26H and I). Erk substrates can compete with the phosphatase for binding to Erk as they use the same docking domain on Erk [151].

In summary, the model fitting has shown that saturation of Mek alone or in

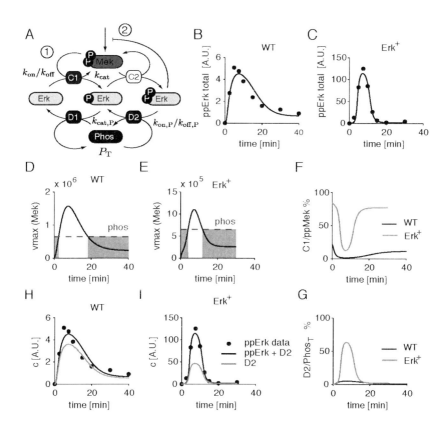

Figure 3.26 Performance of the full model with distributive (de)phosphorylation of Erk. Enzyme-substrate complexes of the kinase (herein called C_1/C_2) and phosphatase (D_1, D_2) that form with the differently modified forms of Erk are explicitly considered (panel A). The model has 7 parameters, including the concentration of the phosphatase P_T. B) and C) show how the best model fit (solid line) compares to the experimental data (black dots). The total amount of ppErk constitutes the sum of the free from and the form which is bound to the phosphatase, given by D_2. The predicted ratio of kinase to phosphatase activity after stimulation is illustrated in D) and E). The areas shaded in grey mark the condition for kinetic suppression of Erk activation. F) and G) show the predicted sequestration of the kinase in C_1 and the phosphatase in D_2. H) and I) correspond to panel B) and C), however they additionally show the predicted time-course of D_2. Model equations in appendix section I.4.

combination with the negative feedback from Erk to Mek is not sufficient to explain the faster downregulation of Erk activity at Erk overexpression. Additionally, some regulation must happen on the level of the phosphatases. Possibly, phosphatase activity is upregulated in a fast and Erk dependent manner. Alternatively, the model of distributive phos - and dephosphorylation has the intrinsic ability to explain the data, due to saturation and sequestration of the participating modifying enzymes.

The "kinetic suppression of ppErk formation" is a steady state phenomenon. It is found to be relevant in all the model fits here, as the kinetics of ppMek, the input to the system, is the slowest variable in the system. Thus, phosphorylation of Erk is in a quasi-steady state with respect to the ppMek signal. Due to the multiple layers of feedback regulation in the MAPK signalling pathway, it is not clear, how the relatively slow response, which extends over more than 30 min here, comes about. However, as measured phosphoErk and phosphoMek1 signals are characterised by the same response kinetics, a swift signal transmission from Mek to Erk seems likely.

Up to now I have shown that the deactivation of Erk proceeds much faster in Erk overexpressing cells and discussed mechanisms that might explain this. As the different Erk phosphorylation kinetics establish rather similar levels of Erk activity roughly 20 min after stimulation, it is tempting to speculate that the width of the ppErk peak constitutes a robust pathway output parameter. If the absolute ppErk levels are irrelevant for signal transmission, downstream targets of Erk should not be affected by fluctuations of the Erk expression level. For this reasons I follow up with an analysis of Erk target activation.

3.6 Increased phosphoErk signals are attenuated at the level of Erk targets

I have analysed two direct cytoplasmic Erk targets, p-p90RSK and p-p70S6K along with ppMek1 and ppErk2 in response to EGF stimulation using the Bioplex assay. Whether the herein measured phospho sites of p70S6K are direct targets of Erk phosphorylation has not been shown in a stringent fashion, however all data are in agreement with this (additional information can be found in the appendix section C.3).

The working principle of the Bioplex system and the experimental procedures

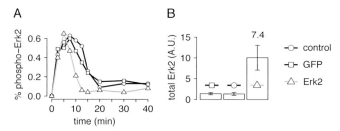

Figure 3.27 EGF time series in GFP and Erk2 overexpressing Hek293 cells. A) Quantification of relative phosphorylation of Erk2 after EGF (25 ng/ml) stimulation from a Western blot in normal Hek293 cells (20 μg lysate, control) and Hek293 cells expressing either GFP (20 μg lysate) or Erk2 and GFP separately (6 μg lysate). B) Quantification of total Erk2 level (normalised to Erk1) in the three different cell lines in A). The number on top of the bar indicates the average fold change of Erk2 expression.

as well as data processing are described in the appendix sections C.1 and C.2. Shortly, proteins of interest are captured by beads which are coated with the respective primary antibody. A secondary, labeled antibody binds to the phosphorylated form of the protein. The signal of the label is interrogated from several (≈ 100) beads per sample and the median of the signal is the final readout. Several phospho-proteins can be analysed at the same time, as every bead can be identified by its unique combination of two internal dyes. Due to the way the assay is built, the measured signal is proportional to the relative phosphorylation of a protein and to compare different samples with each other, it is important that the same amount of protein has been incubated with the beads (for details see section C.1). For this reason, it is not safe to directly compare Erk2 phosphorylation of WT and Erk2 overexpressing cells in this assay. To estimate the degree of Erk overexpression and analyse Erk phosphorylation kinetics in a reliable fashion, I also analysed one biological replicate on a Western blot as described previously (Fig. 3.27). The relative phosphorylation of Erk2 follows the same kinetics in WT Hek293 cells and cells expressing only GFP, which verifies that the coexpression of GFP in the T2A system can be tolerated. Also, it could be verified that the relative phosphorylation of Erk2 declines faster in Erk2$^+$ cells (see Fig. 3.27A), when overexpression of Erk2 was roughly 7 fold (see Fig. 3.27B). The % of phosphorylated Erk2 is identical at the peak in WT and Erk2$^+$ cells about 5 min after stimulation.

In each of the three biological replicates it can be found that phosphorylation of Mek1 is significantly reduced in Erk2$^+$ cells already 2.5–5 min after stimulation (Fig. 3.28), which confirms the action of a fast negative feedback.

Though Erk activity is linearly increasing with the amount of Erk at the peak, activation of p90RSK was only slightly increased in Erk2$^+$ cells so that the higher signal is clearly attenuated on the level of the targets. Interestingly, the sharper decline of Erk2 phosphorylation after maximal stimulation is somewhat visible in the kinetics of the p-p90RSK activity pulse (Fig. 3.28). The p-p90RSK assay has not reached its upper intensity limit, as the raw value of p-p90RSK was even higher in a positive control. That means, the rather identical fold changes of p-p90RSK in WT and Erk2$^+$ cells are no technical artefact. An additional experiment confirmed that even at roughly 20 fold Erk overexpression the fold induction of p-p90RSK remained unperturbed compared to WT cells (see appendix section C.4).

At the target p70S6K, phosphorylation is increased maximally 1.5 fold when comparing Erk2$^+$ cells with WT or GFP only expressing Hek293 cells. Also here the signal from phosphoErk is clearly attenuated. What can be the reasons for signal attenuation at the level of the targets?

The simplest explanation would be saturation. Given that for example under normal circumstances 66% of p70S6K is phosphorylated at the peak (real estimate not known!), an increase to 100% phosphorylation would correspond to a fold change of 1.5. A different explanation would be that unphosphorylated as well as phosphorylated Erk bind tightly to targets - as a consequence only the relative Erk activity ppErk/Erk$_T$ would be transmitted to the next level. Additionally active ppErk molecules might be sequestered with negative regulators like phosphatases, making them inaccessible to their substrates. Last but not least, nuclear translocation of Erk might change its accessibility to targets in the cytoplasm [29]. An increased induction of immediate early genes could be the consequence, however, in a RT-PCR analysis induction of *EGR1* and *c-Fos* m-RNA was unaffected by Erk overexpression (see Fig. 3.29). It has been shown that the induction of *c-Fos* in humanT memory cells can be titrated with various levels of U0126, an inhibitor of Mek1/2 [14]. However, it is possible that without pathway inhibition, induction of transcription is already saturated, such that even higher signals cannot produce higher fold induction of those genes. It seems more likely though, given the fact that the phosphorylation of immediate cytoplasmic Erk targets is already attenuated, that this attenuation is amplified along the cascade. Finally, only the duration of the activity pulse might

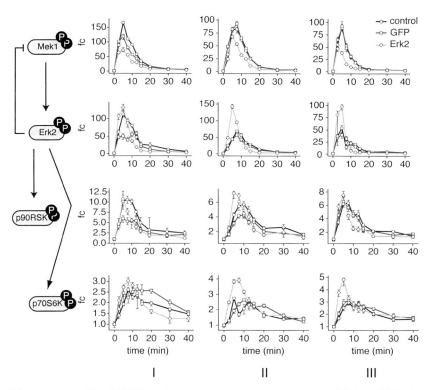

Figure 3.28 The MAPK pathway response at overexpression of Erk2. Multiplex analysis of phosphorylation of Mek1, Erk2, p90RSK and p70S6K after stimulation with EGF (25 ng/ml). Shown is the fold induction relative to the mean of three untreated controls for normal Hek293 cells and cells expressing GFP or Erk2 (and GFP separately) in 3 biological replicates (columns I to III). In replicate nr. I the same lysates as in the Western analysis of Erk phosphorylation shown in Fig. 3.27 have been used. Error bars indicate the 95% confidence interval (see also appendix section C.2).

control gene induction, as speculated earlier. In line with this the fold change of p-p90RSK and p-p70S6K is rather similar 20 min after stimulation, except for one of the three replicates, where the fold induction of p70S6K is even lower in Erk overexpressing cells after 20 min (see p70S6K data in replicate I in Fig. 3.28).

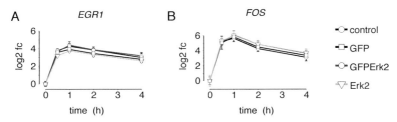

Figure 3.29   RT-PCR analysis of *EGR1* and *c-Fos* m-RNA induction. WT Hek293 cells or cells expressing GFP, GFPErk2 or Erk2 (T2A construct) have been treated with EGF (25 ng/ml) and mRNA was isolated 0.5, 1, 2 and 4 hours after stimulation (protocol in section F). The plots show the mean log2 fold change of mRNA induction (average of 3 biol. replicates) with respect to t=0 for A) *EGR1* and B) *c-Fos*. Erk overexpression was about 7 fold in Hek293 cells expressing Erk2.

3.7 Discussion

Robustness of the inactive pathway state at increased Erk expression levels

In this chapter I have studied how Erk overexpression affects the transient MAPK pathway response to EGF stimulation using theoretical and experimental approaches. The question is interesting not only in the light of robustness, but also with respect to cancer research. It is generally believed that overexpression of pathway components leads to aberrant pathway activation even in absence of a stimulus, as is the case for EGF receptor overexpression. However, the theoretical analysis of a distributive dual phosphorylation cycle shows that phosphorylation of Erk is attenuated at higher expression of Erk and thus predicts robustness of the inactivated pathway state.

General design principles of signalling pathways that affect the stability of the "off state" at spurious and/or weak pathway stimulation have been discussed by Heinrich et al. [64] and concern the specificity of kinases, the position of positive feedback loops or signalling crosstalk. Here my focus was on finding mechanisms that establish concentration robustness. Single phosphorylation cycles possess the ability to attenuate activation of the kinase at increased expression levels [96], however dual phosphorylation cycles amplify this effect and generate a bell-shaped function of Erk activity depending on the Erk expression level. The reason for this non linear relationship is the saturation of the upstream kinase

Mek in combination with sequestration of Mek within the first phosphorylation cycle. The prerequisite for this mechanism is that Mek operates in a distributive fashion and that the kinase Mek has a lower v_{max} than the counteracting phosphatase.

Importantly, the mechanisms described here constitute robustness towards fluctuating levels of only Erk. Networks in which the steady state of a component only depends on kinetic parameters possess ACR, absolute concentration robustness [139]. In a single phosphorylation cycle, pErk reaches a limit for increasing levels of total Erk, however, the steady state still depends on the concentration ratio of the modifying kinase and phosphatase (see eq. (3.1)). Bifunctional enzymes which possess two catalytic sites for kinase or phosphatase activity automatically have a kinase to phosphatase concentration ratio of 1. Thus, if Erk would be regulated by a bifunctional enzyme, the single phosphorylation cycle would possess the property of absolute concentration robustness [146].

Distributive phosphorylation cycles have been extensively studied theoretically and where shown to produce more complex dynamic behaviour than could be expected intuitively. Phosphorylation cycles can implement ultrasensitive, switch like responses and even exhibit bistability. First theoretical findings date back to the 1980s [57], however even nowadays new mechanistic insights are found [131, 132, 18, 95, 103].

Not only because of validity within restricted parameter regimes but especially because of the question as to whether Erk phosphorylation is distributive, the *in vivo* relevance of these phenomena is always a discussion point. In this work unfortunately I did not succeed in clarifying this question. The annotation of spectra from isoelectric focusing that show the differently modified states of Erk1/2 is a challenging and time consuming task that remains to be undertaken. While it is clear that Erk phosphorylation is distributive *in vitro* [63, 48], two mechanisms come to mind that could convert a distributive mechanism to a quasi-processive one *in vivo*.

First, cells are anything but perfectly mixed low-concentration reaction systems - molecular crowding might lead to limited diffusion and thus increase the probability for a recently dissociated Mek-pErk complex to reassociate for the second phosphorylation step. In HeLa cells, molecular crowding has been identified as the reason for a quasi processive phosphorylation of Erk [6, 5]. A second mechanism is an increased stability of the Mek-pErk complex due to anchoring to molecular scaffolds. However, in HeLa cells the contribution of scaffolds was basically excluded, as the knockdown of either of the scaffolds KSR1, MP1, IQ-GAP1, Paxillin, β-Arrestin1 or β-Arrestin2 did not lead to significant changes in the phosphoErk1/2 dynamics [6]. Also, it has been shown for MEFs (mouse embryonic fibroblasts), that only KSR1 and Mek1/2 form a rather stable complex in the cytoplasm, whereas the interaction of the scaffold with Raf and Erk is highly dynamic [105]. In conclusion, the scaffold hypothesis stays disputable.

Surprisingly, another study finds strong arguments for distributive phosphorylation in HeLa cells. When comparing cells of different Erk expression levels, the high expression groups showed increased Erk dual phosphorylation at acute stimulation, however, the onset of phosphorylation was delayed in a way that maximal activation was only reached at later time points. Only the distributive model was able to reconcile these data in Hela and MCF7 cells [122]. Futher indications for distributive phosphorylation have been found for primary erythroid progenitor cells [134]. Seven minutes after stimulation, more pTyrErk1/2 had formed than ppErk1/2, as revealed by quantitative mass spectrometry. The authors conclude distributive phosphorylation as the reason, especially as dephosphorylation by DUSP-6 had been shown to be ordered with preference for phospho-tyrosine [176], so that in case of dephosphorylation as driving process pThrErk1/2 would accumulate instead of pTyrErk1/2. Similar phospho-form patterns were seen in HT29 cells which had been treated with Insulin and TNFα for 30 mins. From Erk1 only 20% was fully activated, whereas single phosphorylated Erk1 dominated with 27% . The disproportion was even more pronounced for Erk2, with 55% pErk2 but only 8% fully active ppErk2 [116]. In another study it was tested which model could explain the observed time-course data of ppMek1/2 and ppErk1/2 in PC12 cells. The four competing models suggested all possible combinations of distributive/processive (de)phosphorylation. A Bayesian approach was used for model selection which favoured distributive phosphorylation, while no preference for a specific dephosphorylation mechanism could be found [153]. Thus, up to now, the evidence for distributive Erk phosphorylation *in vivo* outweigh the evidence for a (quasi)processive mechanism.

Regardless of the cause, experimental data are in accordance with robustness of the inactive pathway state. Hek293 cells that were kept in medium supplemented with 10% FCS showed no detectable levels of ppErk on a Western blot even at 20x overexpression of Erk. The presence of fetal calf serum in the medium can be considered as weak pathway stimulation, as in comparison levels of EGF were measured to be lower than 2pg/ml in 10% FBS (fetal bovine serum) [149], which corresponds to approximately 10000x lower levels of EGF than were used to stimulate (25 ng/ml) Hek293 cells in this study. However, also in other studies where WT Erk was ectopically expressed in Cos-1 cells [43] or CCL39 chinese hamster lung fibroblasts [66, 118], Erk activity towards MBP remained very low in serum starved cells. The FACS analysis however indicates that some ppErk can be found in the Hek293 cells, as the measured signal is clearly shifted compared to the signal of an unspecific control antibody of the same isotype (see appendix section E.3). Also, ppErk has been found to be slightly increased in Erk overexpressing cells – however, total Erk and ppErk were rather uncorrelated in those cells.

The prevalence of Erk overexpression in cancer

The activation of a kinase by distributive phosphorylation of two or more sites could be a general motive which confers robustness to the inactive kinase state. As cancer cells arise from an evolutionary process in which beneficial traits are selected for, the frequency of Erk overexpression in malignant transformation could be an indirect measure of robustness at this level of the pathway.

To compare the frequency of EGF receptor (genes EGFR, ERBB2, ERBB3, ERBB4) and Erk overexpression (genes MAPK1 and MAPK3) I have searched for copy number alterations of those genes using cbioportal [34, 53]. The cbioportal database currently disposes of the data from 20958 samples out of 89 cancer genomics studies. Table 3.1 shows the tumour types with the highest and second highest frequency of amplification of the respective genes. Overexpression of EGFR is found in 43% of glioblastoma (and only 7% of cases of EGFR overexpression coincide with an EGFR mutation), while overexpression of Erk reaches frequencies of around 5% at maximum throughout all tumour types.

In glioblastoma the overexpression of the receptor clearly seems to confer an advantage, which cannot be replaced by amplification of Erk1/2 (see Fig. 3.30). Only in two out of 574 cases Erk was amplified and this is confronted by some other two cases in which one isoform of Erk was deleted. Also, the amplification of receptor and Erk is not mutually exclusive.

A counterexample are sarcoma with 5% MAPK1 amplification and blad-

gene	freq. %	tumour type	# samples	ref.
EGFR	43	glioblastoma	574	[20]
	14	esophageal carcinoma	184	*
ERBB2	15	esophageal carcinoma	184	*
	14	stomach adenocarcinoma	441	*
ERBB3	5	uterine carcinosarcoma	56	*
	5	ovarian serous cystadenocarcinoma	579	*
ERBB4 [1]	5	prostate adenocarcinoma	56	[10]
	4	ovarian serous cystadenocarcinoma	579	*
MAPK1	5	sarcoma	257	*
	5	ovarian serous cystadenocarcinoma	579	*
MAPK3	6	bladder urothelial carcinoma	128	[152]
	5	breast cancer patient xenografts	116	[45]

*   provisional data of the Cancer Genome Atlas Research Network
[1]   deletion of the gene is an equally frequent copy number alteration observed throughout all cancer types

Table 3.1   Cbioportal, a database that provides access to genome scale cancer studies, was queried for copy number alterations of the listed genes throughout all cancer types. The table shows the frequency of gene amplification in the top two tumour types (sorted according to frequency of this particular gene amplification). The number of analysed samples and the reference of the study is also shown. EGFR, ERBB2, ERBB3, ERBB4 constitute the EGF receptor family, MAPK1 corresponds to Erk2, MAPK3 to Erk1.

der urothelial carcinoma with 6% MAPK3 amplification (8% when considering MAPK1 and MAPK3). Here amplification of Erk and the receptor appear as mutually exclusive events. The occurrence of only one or the other type of modification is usually interpreted to show that one of these events is sufficient for achieving the desired trait in cancer. However, this analysis was restricted to only two proteins and it cannot be excluded that in case of receptor and Erk overexpression other mutational hits are required for constitutively activating the pathway. Finally, even in these two tumour types the amplification of the receptor (all 4 genes taken together) is still more frequent.

The prevalence of amplifications is also higher in several other pathway components downstream of the receptor. AKT (AKT1,AKT2 or AKT3) is amplified in 22% of ovarian serous cystadenocarcinoma (579 samples), the PI3K (PIK3CA, PIK3CB, PIK3CD or PIK3CG) in 48% of lung squamous cell carcinoma (501 samples), Myc in 43% and Raf (ARAF,BRAF,RAF1) in 18% of ovarian serous

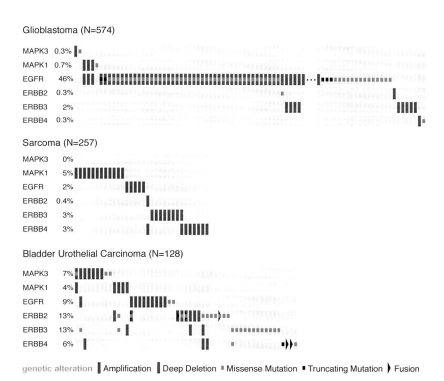

genetic alteration ▌Amplification ▌Deep Deletion ▪ Missense Mutation ▪ Truncating Mutation ❭ Fusion

Figure 3.30 Shown are the so-called oncoprints of the tumour types with the highest frequency of EGFR amplification (glioblastoma), MAPK1 amplification (sarcoma) and MAPK3 amplification (bladder urothelial carcinoma). Each column corresponds to one specific tumour sample, all unaltered cases (samples with no amplification/mutation in EGF receptor or Erk1/2 genes) have been removed from this view. The sarcoma study provided only information on copy number alterations. In the glioblastoma study some cases of EGFR amplification are not shown, as indicated by the black dots. The frequency of EGFR overexpression in combination with an EGFR mutation is 7%. The illustrations were generated with and downloaded from http://www.cbioportal.org.

cystadenocarcinoma (provisional data of the Cancer Genome Atlas Research Network). The activation mechanism of these pathway components differs remarkably from the way Erk is activated by Mek. The PI3K is not activated by phosphorylation, but by direct interactions of its regulatory domain with the phospho–YXXM motif (X indicates any amino acid) of the receptor and direct interactions of its catalytic domain with Ras [37]. AKT requires phosphorylation of Thr308 and Ser474, but obviously these two phosphorylations are carried out by two different kinases, namely PDK-1 and ILK [123], respectively, so that the sequestration of one of these kinases does not effect the operation of the other. Hence, overexpression of these components can be effective in malignant transformation. The transcription factor and oncogene Myc regulates the expression of genes that are involved in cell growth, proliferation and transformation. Its activity level is directly linked to its expression level and mitogenic stimuli effectuate control over its expression and its stability — e.g. active Erk phosphorylates Myc at S62 and thus increases its stability [39]. In consequence amplification of Myc is a simple transformation strategy and the genetic events leading to overexpression have been studied in detail [106]. The activation mechanism of Raf is complicated and multifaceted [129]. However, dimerisation of the different isoforms is crucial for activation and the overexpression might contribute to increased spontaneous dimerisation.

These examples suggest that overexpression might be successful when the activation mechanism of a pathway component relies on direct protein-protein interactions or when the activity is given by the amount of protein itself, as is the case for Myc. However, the fact that Erks' activation mechanism might prevent aberrant activity at increased expression levels of Erk is only one explanation for a lower frequency of Erk amplification in cancer.

An alternative explanation would be that constitutive Erk activity in isolation is either deleterious or insufficient for inducing downstream responses, as the receptor usually activates a whole network of components alongside Erk. In line with this, constitutively activating mutations of Erk are rare. The first reported Erk2 mutation lies within the common docking domain [7]. The mutation was proven to lead to constitutive phosphorylation of Erk, however in a Mek dependent manner. It was shown later that the mutation leads to loss of *in vitro* binding of Erk to DUSP1 [100]. This indicates, that Erk mutations might contribute to increased signal duration, however the signal has to be provided from upstream. Intuitively the same should be true for Mek mutations, as the only targets of Mek are Erk and HSF1 [150]. Indeed the natural occurrence of this pathway component mutation is rare [104, 13]. However, several studies confirm that active mutants of Mek induce oncogenicity in various cell types [27, 101, 104].

Single cell and bulk analysis shows that Erk activity is downregulated much faster in Erk overexpressing cells

The transient Erk pathway response to EGF of Hek293 cells was studied at normal and elevated levels of Erk2. The Western blot analysis of total Erk and phosphorylated Erk shows that Erk activity correlates with Erk expression level at the time of the maximal pathway response, around 5 to 7.5 min after stimulation with EGF. However, Erk activity is downregulated much faster in Erk overexpressing cells so that levels of ppErk are of the same order of magnitude in WT and Erk2^{+} cells already 20 min after stimulation. This indicates the action of a fast negative feedback. Also in the single cell analysis the correlation of Erk expression level and Erk phosphorylation was characterised by the highest Pearson correlation around the time of the maximal ppErk response.

Interestingly, the single cell analysis shows that the levels of ppErk distribute in a bimodal fashion during deactivation 10 to 20 min after stimulation in Erk overexpressing cells. These two populations did not arise due to different Erk expression levels. Somehow the deactivation rate differs between those cells, but why this difference should create two populations instead of a continuous distribution of different phosphorylation states is not clear. More likely, as the ppErk distribution 10 min after stimulation is aligned with the distribution in the untreated state, the bimodal distribution at later time points could represent a damped oscillation. Only an analysis where phosphorylation of Erk is followed over time within a single cell could proof this. Oscillations of ppErk have been observed and are commonly attributed to negative feedbacks [136, 1]. Last but not least the two populations could be some technical artefact - repeating the experiment and optimising for a higher cell count would be desirable.

Mechanisms that explain faster activity decline of Erk at increased Erk levels

Mathematical modelling was used to test which mechanisms might explain faster activity decline in Erk overexpressing cells. The presence of a negative feedback on ppMek1 could be confirmed also experimentally. However, distributive kinetics of Erk activation in combination with the negative feedback on ppMek were not sufficient to explain the data. Some effects have to occur at the level of the phosphatases. A model in which the dephosphorylation rate constants are different in WT and $Erk2^+$ cells explains the data. At first it was surprising that the model predicts a lower dephosphorylation rate in Erk overexpressing cells. However, this is because in this model low levels of ppErk at later time points are due to saturation and sequestration of Mek, while the very high ppErk values around 5 min after treatment can only be achieved if the counteracting phosphatase activity is relatively low. This suggests that two scenarios are possible: either the dephosphorylation rate is upregulated in a fast Erk dependent manner or the phosphatase activity is restricted at the peak of Erk activity. This might be a consequence of phosphatase saturation, as confirmed by a model fit.

With respect to the possibility of fast mechanisms of phosphatase regulation, not much can be found in the literature. Transcriptional feedback leads to newly synthesised mRNA of DUSPs already 30 min after stimulation, however, this mRNA has to be translated to proteins first, which makes this feedback to slow to be of relevance. Also the reported stabilisation of MKP1 by direct phosphorylation by Erk [22] requires the synthesis of new protein to significantly increase the level of active phosphatase. The phosphatase PP2A dephosphorylates Erk and Mek [72] and its expression level is kept in homoeostasis in NIH3T3 and HeLa cells by regulatory events at the level of translation [12]. However, the activity of the catalytic subunit of PP2A can be regulated by the interaction with different regulatory B-subunits. CK2α acts as a regulatory B-subunit for the core PP2A [91] as this kinase has been shown to interact with and to phosphorylate PP2A to in turn increase its catalytic activity towards Mek1 [67]. The expression of CK2α also decreased Erk activity in proportion to the amount of CK2α already 10 min after stimulation in NIH3T3 cells [91]. If CK2α is activated by mitogenic stimuli, this pathway could constitute a fast negative feedback on Erk activity. Another indication comes from a study in human lung fibroblasts, where the PP2A activity was influenced by Erk. However the pathway and most importantly the time scale of this interaction is not known [11]. Thus it remains unclear whether fast post-translational modification of a phosphatase is involved and if so, which phosphatase would be in charge.

The alternative explanation would imply that phosphatase regulation is not required at all. As described earlier, the formation of dual phosphorylated Erk is restricted when active Mek is sequestered in a complex with unphosphorylated Erk. However, at maximal pathway stimulation the v_{max} of Mek might exceed the catalytic rate of the phosphatase, which swaps the modification cycle towards saturation and sequestration of the phosphatase. This phosphatase saturation then facilitates the activity of Mek towards its substrate Erk and enables phosphorylation proportional to the level of Erk. However, as soon as Mek activity has sufficiently declined the phosphatase is released from saturation and causes the seemingly faster dephosphorylation of Erk. This possibility is intriguing, as it does not require any additional layers of regulation.

The prerequisite for the described two mechanisms is the quasi steady state dynamics of Erk with respect to its modifying enzymes. In the first place this is owed to the way the models were formulated, using the measured ppMek1 signal as input. However, the ppMek response is cleary shaped by processes with different time scales. In WT and Erk2$^+$ cells ppMek1 has the same slow time scale on which it peaks and goes back to basal levels. The negative Erk dependent feedback only changes the amplitude of this response, which could indicate that it works on a much faster time scale. This suggests that Erk phosphorylation might really be in quasi steady state with respect to the ppMek input. The slow component of the pathway response might be a result of advancing receptor internalisation by endocytosis, as 90% of the receptors are internalised 20 min after stimulation with 20 ng/ml EGF in murine fibroblasts [145]. This is in agreement with an EGFR endocytosis rate of 0.13 min^{-1} reported previously [162]. On the other hand, signalling has been reported to continue from intracellular compartments like endosomes [111, 107] and measured receptor half lifes of about 45 min exceed the duration of the response measured herein [145]. In line with this, the transient MAPK activation kinetics where not changed even after impairment of clathrin-mediated endocytosis [145]. In conclusion, the contribution of the various feedbacks to the response kinetics remains obscure.

The mechanisms considered in the mathematical models are the most simple ones one can come up with. What has been ignored throughout the investigations are the local aspects of MAPK signalling. Scaffolds and other localised interaction partners may protect ppErk against dephosphorylation, as many phosphatases and substrates use their D-domains to bind to the common docking domain of Erk [142, 151]. When Erk is overexpressed, the scaffolds might saturate and the free ppErk might get dephosphorylated much faster. Another scenario that can arise from the saturation of cytoplasmic interaction partners is the increased translocation of Erk to the nucleus. Then the dephosphorylation rates might simply differ in WT and Erk2$^+$ cells as different phosphatases are localised to the nucleus and cytoplasm. Thus, to solve the riddle of increased dephosphorylation rates at Erk overexpression, it will not only be necessary to investigate the activity of phosphatases, but also the cellular localisation of Erk.

The pulse length of the transient response could be a robust pathway feature

Looking at the response of ppMek1, ppErk2 and the phosphorylation of Erk targets together, signal duration seems to increase slightly from layer to layer, however independent of Erk expression level. Within a layer only the amplitude of the pulse is modulated. Experimental data also show that the increased ppErk signal at Erk overexpression is attenuated at the level of Erk targets, which suggests that signal duration might be the robust pathway feature in contrast to the amplitude of the signal.

This hypothesis is in agreement with other studies on quantitative Erk signalling. In the study of Albeck et al. Erk signalling and the induction of proliferation have been analysed in MCF-10A mammary epithelial cells on the single cell level [1]. At low levels of EGF, single cells showed stochastic pulses of Erk phosphorylation, which increased in frequency and duration at increasing levels of EGF. However, these oscillations did not seem to transmit any additional information, as the commitment to S-phase was linked to the total time spent in the "on state", hence the absolute duration of the signal. Signal duration was also found to be at the basis of the cell fate decision in PC12 cells and the network logic behind this has been explained [133]. In another PC12 cell study it was shown that while the intensity of the transmitted signal is greatly perturbed by targeted inhibition of pathway components, the mutual information[1] between growth factor and downstream targets was conserved [156]. Another indication

---

[1]The mutual information is the logarithm of the average number of discrete signalling states that can be resolved. In this study it was also found that PC12 cells are only capable of binary decisions.

for a high importance of signalling timing is the fact that the transcriptional target c-Fos is rapidly degraded if not stabilised by direct phosphorylation by Erk, which requires the Erk activity to be sustained over the period of a transcription and translation round [112]. A different study suggested that the fold change in nuclear Erk after stimulation was robust in the face of natural Erk expression level fluctuations [35]. However, this would only explain robustness for nuclear Erk targets. Interestingly the authors also write that the Erk dynamics was more reproducible between cells than the amplitude.

Finally, a next step might be to not look for robust features in the ppErk signal alone, as the activated EGFR relays the signal to at least 6 distinct biochemical pathways via the multitude of different adaptors and enzymes that interact with it [9]. Coming back to the example of PC12 cells, it was shown that sustained Erk activity is associated with differentiation, however it was also shown that Raf/Mek/Erk pathway activity in isolation is not sufficient to induce differentiation [157]. The ultimate response might be the result of various outputs from the signalling network that are processed via AND/OR gates. This process could explain how graded responses are converted to all or none responses and would create robustness at the same time. The use of a signalling network instead of a single pathway also offers the possibility for compensation, where alternative pathway routes can be used when some of the nodes have been perturbed [156]. Also, building a signalling cascade with more layers than one would consider necessary might create robustness: if levels of signalling pathway components fluctuate randomly and independently of one another, eventually higher levels of one phospho-protein might be attenuated in another layer of the cascade where lower levels of the protein than usual are expressed.

4 Conclusion

It is believed that robustness is an inherent property of complex evolvable systems. Robustness may facilitate evolution and evolution selects for robust features [83]. Gene expression noise is one of the key perturbations that cellular function has to withstand [140].

To uncover robustness mechanisms in MAPK signalling, expression levels of the terminal kinase Erk have been altered by targeted knockdowns and overexpression. A negative feedback from Erk to c-Raf was shown to upregulate Mek activity and in turn the activity of Erk to compensate for the partial loss of Erk at targeted knockdowns. Supported by a mathematical model these results suggest that negative feedback regulation is an important pathway feature for creating robustness. Additionally, the results suggest redundancy of the pathway at the level of Erk, as a reduction in Erk1 level leads to a compensatory upregulation of ppErk2, and vice versa. The deletion of one isoform of Erk can also be found in cancer cells (see Fig. 3.30).

The partial robustness of absolute ppErk levels was found in colon cancer cell lines with constitutive pathway activity, which restricts conclusions about the actual robust pathway feature. I confirmed the contribution of a fast negative feedback acting upstream of Mek during transient pathway activity in Hek293 cells — the amplitude of the ppMek1 signal was found to decrease as early as 5 min after stimulation in Erk overexpressing cells. However, this feedback cannot prevent Erk phosphorylation from increasing linearly with the amount of Erk by the time of maximal pathway activity. Though the amplitude of the transient phosphoErk signal is increased in Erk overexpressing cells, phosphorylation of Erk is reduced with much faster kinetics. Mathematical modelling predicts that when Erk is activated by Mek in a distributive fashion, the dual modification cycle intrinsically explains these dynamics. Alternatively faster Erk deactivation at higher Erk expression levels could be explained by a fast Erk dependent activation of a phosphatase.

Activation of a kinase by distributive multisite phosphorylation has the potential to be a general motif for concentration robustness of the inactive kinase state. That is why further investigations should be directed to clarifying the *in vivo* mode of Erk activation. Versatile kinetics can result from distributive phosphorylation reactions, however, their significance may be remote if Erk activation is found to be quasi-processive. To identify the mechanism in action that causes fast Erk-dependent Erk deactivation it will also be necessary to investigate the relative cellular localisation of Erk at different Erk expression levels. Experiments in which specific phosphatases are inhibited might also be helpful.

Another important finding of this study is that transiently increased phospho-Erk signals (caused by increased levels of Erk) are attenuated at the level of the Erk targets p90RSK and p70S6K. The reason for this is unclear. Possible mechanisms are the sequestration of ppErk in substrate-inaccessible complexes or the tight docking of inactive and active Erk forms to substrates. Here mathematical modelling could prove useful to test different hypotheses. Of course some of this signal attenuation is due to artificial Erk overexpression: the extraordinary amounts of ppErk can only form due to Erk overexpression, which means that similar fold changes in target activity are technically impossible. To resolve signal attenuation due to saturation, experiments in which Erk and a target protein are jointly overexpressed would give insight. Additionally the amount of ppErk might not be the only indicator of Erk activity; other undiscovered inhibitory phosphorylation sites might contribute to Erk regulation. The complex spectra of the different modification states of Erk obtained from isoelectric focusing suggest that other post-translational modifications of Erk can be involved.

The induction of the immediate early genes *EGR1* and *c-Fos* was also uninfluenced by Erk overexpression. Also here saturation might be the cause and gene induction alone might not be the appropriate readout for the nuclear activity of Erk. To determine the cytoplasm/nucleus activity ratio of Erk, it would be good to jointly measure cytoplasmic and nuclear phosphorylation targets of Erk.

In contrast to the amplitude of the signal, the timing of Erk and target phosphorylation is rather similar in WT and Erk overexpressing cells, which suggests that signal duration might be a robust pathway feature. However, maybe the robust feature is not found in the Erk signal, but within the joint output of the signalling network. A network can contribute additional robustness e.g. by providing alternative pathway routes.

The robustness mechanisms found here will affect how the signal transduction network reacts upon targeted inhibition. The effectiveness of Mek-inhibitors used in cancer therapy is compromised by the presence of negative feedback regulation from Erk to c-Raf, however patients with a feedback-impairing B-RafV600E mutation benefit from this targeted intervention. Analysing robustness of other signalling pathways in a similar way will be the key to devise efficient targeted interventions for these and will unveil which mutations in the pathway will break robustness and thereby open the door for efficient intervention.

5 Appendix

A Cells and constructs

Cell culture

Hek293, RKO and HT29 cells were obtained from ATCC (American Type Culture Collection, UK). Hek293T cells were kindly provided by Achim Kramer (lab for chronobiology, Charité Universitätsmedizin, Berlin). Hek293, Hek293T and RKO cells were kept in DMEM (Dulbecco's Modified Eagle's Medium, Lonza) supplemented with 10% fetal calf serum, 1% ultraglutamine and 1% penicillin/streptomycin. The growth medium for Hek293T cells was complemented with 0.8% glucose. HT29 were cultured in L15 medium (Leibovitz's Medium, Lonza) supplemented with 0.1% (w/v) $NaHCO_3$, 4% (w/v) glucose, 0.2% of 1.4 mg/ml insulin, 1% of 10mM MEM vitamins (Biochrom AG), 0.5% of fetuin/transferrin solution (0.12%/0.05%) (w/v), 10% fetal calf serum, 0.5% ultraglutamine and 1% penicillin/streptomycin. All cells were incubated in a humidified atmosphere of 5% CO_2 in air at 37°C.

Cloning of GFP-Erk2

For cloning of the GFP-Erk2 fusion gene I used the Gateway®Recombination Cloning Technology (Life Technologies). The entry clone pENTR-MAPK1 was purchased from Source Bioscience. The entry clone harbours the *Homo sapiens* MAPK1 gene flanked by the recombination sites *att*L1 and *att*L2, a Kanamycin resistance and a pUC origin for efficient replication in E.coli. The lentiviral destination vector containing the sequence of green fluorescent protein EGFP from *Aequorea victoria* in the N-terminal position was kindly provided by Dr. Guido Hermey, Centre for Molecular Neurobiology, Hamburg.

Cloning of GFP-T2A-Erk2

The expression vector for GFP-Erk2 contains a 42 bp linker sequence that connects GFP with Erk2. This linker sequence was replaced with the 63 bp T2A sequence by a fusion PCR. (The T2A sequence is CTT GAG GGC AGA GGA AGT CTT CTA ACA TGC GGT GAC GTG GAG GAG AAT CCC GGC CCT TCC GGT without the initial CTT which was added to create a HindIII restriction site for selection of clones). Two DNA fragments can be fused together

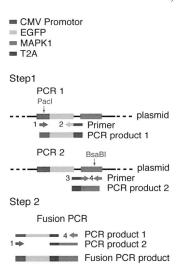

Figure A.1 Cloning of the GFP-T2A-Erk2 construct. A Fusion PCR was used to replace the 42 bp linker sequence between EGFP and MAPK1 (=Erk2) in the GFP-Erk2 construct with the sequence of T2A. The different primers used during cloning (labelled with the numbers 1 to 4) are depicted as arrows coloured according to the sequence they bind to. Step 1) In two separate PCRs the sequence left and right adjacent to the linker is copied and extended by the sequence of T2A by using primers with overhanging ends that contain the sequence of T2A in forward (primer 3) and reverse direction (primer 2). Step 2) In the fusion PCR the previous PCR products are used as templates and align due to the overlapping T2A sequence. In consequence the fusion product is amplified. Unique restriction sites used in this cloning procedure are indicated with red arrows.

in a PCR reaction, when these fragments contain an overlapping sequence of sufficient length. In PCR1 and PCR2 sequences upstream and downstream of the linker sequence are copied (step 1 in Fig. A.1). The copied fragments have to contain unique restriction sites, so that the final insert can be integrated into the plasmid. Primer 1 and 4 were placed so that PCR product 1 contains the restriction site of BsaBI and PCR product 2 the restriction site of PacI. Additionally the reverse primer in PCR1 (primer2) and the forward primer in PCR2 (primer3) contain the T2A sequence as an overhang, which is not complementary to the plasmid.

All PCRs have been performed using Phusion®High-Fidelity DNA Polymerase (NEB) according to the instructions of the manufacturer. The PCR conditions had to be optimised to allow specific amplification of the GC-rich repetitive part of Erk2. Raising the annealing temperature from 60 to 65°C and adding 3% of DMSO led to success. In the fusion PCR (step 2 in Fig. A.1) the product of PCR1 and PCR2 align and the fused fragment is amplified. The original GF-PErk2 expression plasmid and the fusion fragment were digested with BsaBI and PacI and purified using agarose gel electrophoresis and the QIAquick® Gel extraction kit according to the instructions of the manufacturer. Ligation of the fragments was carried out using the T4 DNA Ligase kit from NEB according to instructions of the manufacturer. Suitable clones were preselected by digestion with HindIII, as the GFP-T2A-Erk2 containing plasmid has one additional HindIII restriction site compared to the plasmid containing GFP-Erk2.

I also cloned a plasmid with the GFP-T2Arev-Erk2 sequence, where T2Arev stands for the reverse complement sequence of T2A. The reverse complement sequence contains a stop codon and thus transfected cells will express only EGFP with a shortened T2A tag. Cells transfected with this plasmid have been used as a control.

Stable transfection of Hek293 cells

The GFP-T2A-Erk2 (GFP-T2Arev-Erk2) and GFP-Erk2 constructs have been cloned into lentiviral expression vectors. Cotransfection of Hek293T cells with the expression plasmids and the respective helper plasmids (ViraPower Lentiviral Expression Systems, Invitrogen) leads to the production of replication - incompetent recombinant HIV-1-based lentiviruses. $0.8 \cdot 10^6$ Hek293T cells were seeded in 6cm dishes (medium without antibiotics!) one day before transfection with 4 μg expression plasmid, 2.4 μg pLP1, 1.2 μg pLP2 and 0.8 μg pLP (=VSVG). Transfection was performed using Lipofectamine 2000 according to the instructions of the manufacturer (Invitrogen). The viruses are released to the growth medium which is collected 24h after transfection and snap frozen in liquid N_2 (200 μl aliquots).

Hek293 cells were seeded at $3 \cdot 10^5$ cells/well in a 6 well culture dish one day before viral transduction. Medium containing virus was diluted at 1:10, 1:7.5 or 1:5 in DMEM and added to the cells for 24 hours. Another day later cells showed green fluorescence and cell populations with high transduction efficiency were chosen for further cultivation. Single cell measurements of the GFP label indicated that increased amounts of virus at transduction mainly increased the % of transfected cells, rather than increasing the mean expression level of the transduced gene.

B Semiquantitative Immunoblots

B.1 Protocol

Differently transfected Hek293 cells were seeded at $5 \cdot 10^4$ cells/well in a 12 well culture dish two days before the experiment. Cells were treated with 25 ng/ml EGF and stimulation was aborted by fast medium removal, cooling the cells on ice and subsequent lysis of the cells using cell lysis buffer from Biorad. Reagents for SDS-polyacrylamide gelelectrophoresis (PAGE) and Western blotting were obtained from Bio-Rad Laboratories (Richmond, CA, USA) and Carl Roth (Karlsruhe, Germany). Electrophoresis was performed and lysates were transferred onto nitrocellulose membranes (Schleicher & Schüll). Unbound protein sites were blocked with 1:1 Li-COR buffer in phosphate-buffered saline. Thereafter, specific proteins were detected by incubation with primary antibodies diluted in 1:1 Li-COR buffer/phosphate-buffered saline containing 0.1% Tween-20 (PBST) overnight at 4 °C followed by near-infrared secondary antibodies. The following antibodies were used: mouse anti-human P-Erk1/2 (phospho-p44/42 MAPK Thr202/Tyr204, Cell Signaling Technology, 1:500), rabbit anti-human Erk1/2 (p44/42-MAPK, Cell Signaling Technology, 1:500), rabbit anti-human β-Tubulin (Cell Signaling Technology, 1:2000) or mouse anti-human GAPDH (Ambion, $1:10^4$). Secondary anti-mouse and anti-rabbit antibodies labelled with different infrared dyes (Li-COR) were used to detect signals from different antibodies simultaneously. Membranes were scanned using Li-COR Odyssey.

The bands were quantified using the software Image Studio Lite. The median background fluorescence was measured locally from the boundary regions of each band. The quantified signal corresponds to the sum of all pixel intensities within the band after subtraction of the quantity "number of pixels x median background signal". All band intensities were normalised to a control (Tubulin or GAPDH).

Band intensities from different Western blots are not comparable to each other, even if the data have been corrected for the loading error by normalising to a

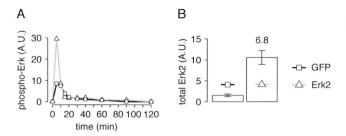

Figure B.1 EGF time series in GFP and Erk2 overexpressing Hek293 cells. A) Quantification of total phosphoErk (phosphoErk1+phosphoErk2) as detected by the ppErk1/2 antibody after stimulation with EGF in Hek293 cells expressing GFP (20 μg lysate) or Erk2 (20 μg lysate). B) Quantification of total Erk2 (normalised to Erk1). The number on top of the bar indicates the average fold change of Erk2 expression.

loading control. However, the measured intensity ratio of two bands which have been detected by the same antibody does compare between different Western blots. To quantify average Erk2 overexpression of a cell type, Erk2 band intensities were normalised to the respective Erk1 band intensities, and then averaged for the 9 (or 10) lanes of each Western blot to obtain mean and standard deviation. The same procedure was taken for GFPErk2 band intensities.

To allow the direct comparison of band intensities from different Western blots, a control lysate was analysed on one lane of each of the single Western blots. The quantified band intensities of the control lysate (normalised to the loading control) were used to normalise the quantification of the different blots with respect to each other.

B.2 Long time series of phosphoErk after treatment with EGF

Differently transfected Hek293 cells have been stimulated with EGF and phosphorylation of Erk was measured from a westerblot 5, 10, 15, 20, 30, 40, 60, 90 and 120 min thereafter. Quantified levels of Erk and phosphoErk are shown in Fig. B.1. This analysis shows that it takes at least 60 min for Erk phosphorylation to drop back to basal levels.

C Bioplex Assay

C.1 The measurement principle

The Bio-Plex 200 Protein Array system (BioRad, Hercules, CA) allows the quantification of up to 100 different phosphorylated proteins. Each assay contains beads whose surface is loaded with antibodies against a specific protein. The bead captures the specific protein and a second biotin-labelled antibody detects only the phosphorylated form. Finally, a fluorescently labelled streptavidin binds to biotin. The integrated fluorescence intensity of all reporters on the bead surface is measured bead by bead in a flow chamber, just like in a regular flow cytometer. To allow the analysis of several different phospho-proteins the beads are internally labelled by two fluorescent dyes. Measuring the fluorescence intensities of these two internal dyes in parallel to reporter interrogation allows identification of a specific bead (xMAP technology) and thus identification of the measured phospho-protein. Some of the experiments have been performed with the MAGPIX system (Life Technologies). Here magnetic beads (identifiable by xMAP technology) are captured and their image is recorded and quantified automatically.

Because of the way the assay is built, the measured reporter values do not reflect the total amount of phosphorylated protein, but rather a mixture of the relative and absolute amount. The assay procedure has been optimised in our lab to produce linearity of the signal with the amount of phospho-protein. However, the relative saturation of the beads is unknown and influences the measured signal as described in Fig. C.1. For that reason, phosphoErk2 signals cannot be compared between WT Hek293 cells and Hek293 cells that overexpress Erk2.

C.2 Protocol and data processing

The protein lysates for Immunoblot analysis have been prepared using the cell lysis buffer from Biorad, which is also recommended for sample preparation

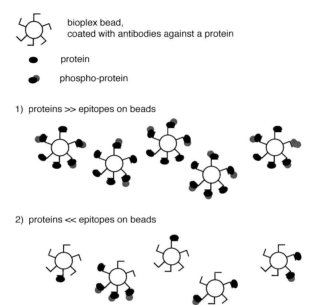

Figure C.1 Measurement of phospho-proteins in the bioplex assay. When 50% of the protein of interest is phosphorylated and there are much more proteins available than epitopes on the surface of all beads in the assay (case 1) all beads are saturated with protein and the signal (proportional to the amount of bound phosphorylated proteins) faithfully reflects the relative phosphorylation. In this regime the signal increases linearly with relative phosphorylation and higher protein concentrations (for example because of overexpression) should not manipulate the final signal.

In the case where there are less proteins than epiptopes, the beads are not saturated and the relative occupation of the beads might vary, which can cause wider distributions in the signal and also a systematically lower signal per bead than one would expect at this degree of phosphorylation (when comparing to case 1). In this setting the signal also increases linearly with increasing relative phosphorylation of the protein. However, when the concentration of total protein increases, this will lead to higher signals as well. This makes lysates of normal cells and cells which overexpress the protein of interest incomparable.

in the Bioplex assay. In most experiments the same lysates have been used for Immunoblot and Bioplex analysis. Lysate protein concentration was determined with the BCA (bicinchoninic acid) method. The beads specific for pp-Mek1 (Ser^{217}/Ser^{221}), pAKT (Ser^{473}), p-p90RSK (Ser^{380}), ppErk1/2 (Thr$^{185/202}$ or Tyr$^{187/204}$) and p-p70S6 Kinase (Thr^{421}/Ser^{424}) as well as the detection antibodies were diluted 1:5. The amount of protein was 15 μg per assay. For acquiring data the Bio-Plex Manager software was used according to the manufacturers instructions. The software makes sure that 100 beads of each type are

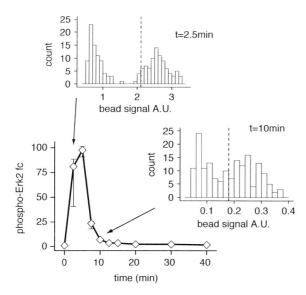

Figure C.2 Processing of bioplex raw data. For each time point the signal intensity is measured of at least 100 beads. The bead signal distribution is shown for t=2.5 and t=10 min. The median is used to quantify the signal and is indicated with the dashed vertical line. The 95% confidence interval of the median is estimated by bootstrapping and used for the error bars in the phosphoErk2 time series shown here.

measured before the machine proceeds to the next sample. However, sometimes formation of bead aggregates is excessive and the count of single beads is reduced. To avoid bias that is introduced by the estimation of median fluorescence from varying numbers of beads, I have pooled the bead reporter fluorescence data from several (up to three) technical replicates in one sample. The median was estimated from the pooled data. Bootstrapping of the sample (n=10000) was used to estimate the 95% confidence interval of the median using the BCA method[44] as implemented in R (R project for statistical computing). Figure C.2 shows an EGF time series of phosphoErk2 where the error bars indicate the confidence interval of the median bead fluorescence. In some instances, the confidence interval is large (at t=2.5 min in Fig. C.2), which is linked to a bimodal bead fluorescence distribution. One of the two intensity peaks localises at fluorescence intensities close to zero, which might indicate the presence of beads with incomplete antibody coating.

C.3 Direct phosphorylation of p70S6K by Erk

Whether p70S6K is a direct target of Erk is still controversial. Whereas the database phospho.ELM lists the herein measured phosphorylation sites Thr^{421}/Ser^{424} as controlled by CDK1 and mTor, the database PhosphoSitePlus only states that they are influenced by treatment with EGF. However, Thr^{421}/Ser^{424} lie within a consensus phosphorylation motif for MAPK (consensus motif is Ser/Thr-Pro with preference for Pro at the -2 position [59]) and p70S6K was found in immunoprecipitates of MAPK and vice versa, MAPK in immunoprecipitates of p70S6K [97]. *In vitro* kinase assays also exclude p38 and JNK as kinases of these sites [175]. It is known, that incubation of deactivated p70S6K with Erk leads to phosphorylation of p70S6K, however the reactivation of the kinase is only partial [110, 175], as well as inhibition of p70S6K is only partial at inhibition of Mek [97]. An analysis of various mutant forms of p70S6K suggests that Thr^{421}/Ser^{424} reside within an autoinhibitory domain. Phosphorylation of these sites by Erk may release autoinhibition and facilitate phosphorylation of Thr^{389} by mTor. Together these phosphorylation events synergies to allow access of PDK1 which fully activates the p70S6K by phosphorylating Thr^{229} [125, 38]. This model can embrace all experimental data and it seems plausible to assume that the p70S6K sites Thr^{421}/Ser^{424} are directly phosphorylated by Erk.

C.4 The MAPK pathway response at 20x Erk2 overexpression

In the main text I show the phosphoMek1 signal in WT and Erk2 overexpressing cells, which was later also used for model fitting (see Fig. 3.21). However, this bioplex assay also covered the analytes ppErk2 and p-p90RSK, which are shown in Fig. C.3.

As described earlier (in section C.1), the phosphoErk2 signal of WT and Erk$^+$ cells is not comparable in the Bioplex assay. In replicates I and II the phospho-Erk2 signal at the peak is similar in WT and Erk$^+$ cells, in accordance with the

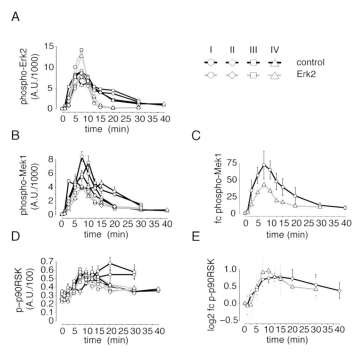

Figure C.3 The MAPK pathway response as measured by the Bioplex assay. Protein lysates have been subjected to a multiplex analysis of phosphorylation of Erk2, Mek1 and p90RSK. Biological replicates III and IV have been obtained from the same lysates as the Immunoblot analysis in Fig. 3.15. A), B) and D) show phosphorylation of the three analytes in a time series after EGF (25 ng/ml) stimulation in four biological replicates of Hek293 cells with and without overexpression of Erk2. Error bars indicate the 95% confidence interval. C) Average in phosphoMek1 signal of the four biological replicates, shown as fold change with respect to untreated control. Error bars indicate the standard deviation of the fold change. E) Average log2 fold change of p90RSK phosphorylation.

assay showing relative phosphorylation (Fig. C.3A). In contrast, replicates III
and IV show increased signals in Erk$^+$ cells, and it is not clear whether this is
due to increased bead saturation or due to actual increased relative phosphorylation. For this reason phosphoErk2 kinetics were analysed on a Western blot
(see Fig. 3.15).

The relative phosphorylation kinetics and intensity of p90RSK looks rather
similar in WT and Erk2$^+$ cells (see Fig. C.3D for absolute measured values and
C.3E for the mean fold change response) even at 20x Erk overexpression. However, the fold induction of p90RSK phosphorylation (compared to the positive
control) is rather weak in this experiment.

D Isoelectric Focusing

Proteins lysates of Hek293 cells were prepared using Bicine/CHAPS buffer according to the protocol of the manufacturer (protein simple™). Proteins were separated by charge on the NanoPro 100 system from proteinsimple™. The sample is loaded to a capillary together with labelled standard peptides of defined isoelectric point (see Fig. D.1). After isoelectric focusing, the sample is immobilised via a proprietary, photo-activated capture chemistry that links the proteins to the wall of the capillary. Then the primary and the HRP-conjugated secondary antibody are loaded one after another. A CCD camera records chemiluminescence of the whole capillary as the HRP (horseradish peroxidase) substrate is flushed through. All consumables as well as primary antibodies specific for Erk1/2 (040-474), Erk1 (040-475) and ppErk1/2 (040-477) and the Goat-Anti-Rabbit secondary antibody (040-656) were purchased from proteinsimple™.

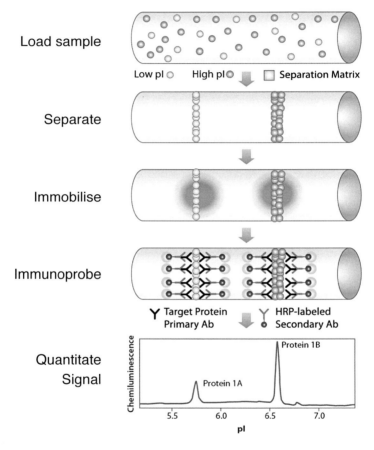

Figure D.1 The principle of capillary-based isoelectric focusing. The figure was kindly provided by proteinsimple™.

E Flow Cytometry

E.1 Protocol and data processing

Hek293 cells were detached using Trypsin or Accutase (Life Technologies) to create a single cell suspension. Aliquots of $4 \cdot 10^5$ cells were prepared for each sample. The single cell suspensions were rested for 30 min at 37°C and 5% CO_2 to avoid influence of mechanical MAPK pathway activation. Cells were stimulated with 25 ng/ml EGF and stimulation was terminated by adding an equal volume of prewarmed (37°C) 4% PFA/BSA or BD Cytofix. Fixation was allowed to proceed for at least 10 min . The cells were washed with PBS/1%BSA solution and permeabilised with ice-cold methanol for 30 min on ice. For immunostaining of the cells antibodies were diluted in PBS/1%BSA. The following primary antibodies were used: rabbit anti-human Erk1/2 (p44/42 MAPK antibody #9102 Cell Signaling Technology, 1:50), mouse anti-human Alexa Fluor 647 anti-pERK1/2 (pT202/pY204) (BD Phosflow, 1:10), Alexa Fluor 647 mouse IgG1 κ isotype control (BD Phosflow, 1:10). For colour detection of total Erk1/2 the following secondary antibodies were used: Donkey anti-rabbit IgG PE (eBioscience, 1:50) or Alexa Fluor 488 goat anti-rabbit IgG (H+L) (Life Technologies, 1:200).

All single-cell fluorescence measurements were performed using the BD Accuri ™C6 flow cytometer. It is equipped with two lasers of 488nm and 640nm wavelength and enables the simultaneous measurement of 4 fluorescence parameters. At maximum three colour-experiments have been made using the labels GFP/Alexa488 (filter 533/30nm), PE (phycoerythrin) (filter 585/40nm) and Alexa647 (filter 675/25nm). Cell sorting according to GFP label intensity was performed using a FACSAria Cell Sorter (BD, excitation 488nm).

Data processing involved 3 steps, 1) gating for viable, single cells, 2) fluorescence compensation (if necessary), 3) regression of fluorescence signals over the scatter parameters FSC-A and SSC-A to remove the influence of cell morphology onto the measured signal. As a first step of data processing the measurements

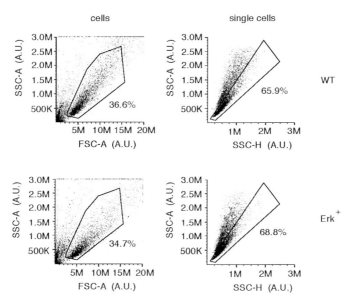

Figure E.1 Gating for viable, single Hek293 cells. The single cell data were at first gated for viable cells, using the SSC-A (side scatter, area of pulse) and FSC-A (forward scatter, area of pulse) parameters which reflect granularity and size of the cells. A second gate was set to eliminate cell doublets. The cytometer measures height and area of each pulse. Cell doublets generate pulses with the same height as single cells, however, as a cell doublet passes through, the pulse has an increased length and in consequence, an increased area. Cell doublets appear as a shadow population with increased area in the SSC-A vs. SSC-H scatter plot. The number next to the polygon gates indicates the % of events within the gate. Within the scales K stands for 10^3 and M for 10^6.

have been gated for viable, single cells as illustrated and described in Fig. E.1. All Erk2 overexpressing cells also express GFP, so that total Erk1/2 was detected using an antibody with PE label in these cells. As GFP and PE show significant spectral overlap, compensation of fluorescence spillover from one parameter to the other is necessary. Traditionally, fluorescence compensation was set before the measurement. However, as compensation is a digital process, it can be carried out even after data acquisition. The advantage is that errors introduced by wrong compensation can be corrected anytime [68]. Here compensation has been carried out after data collection and proper gating of single viable cells using the compensation tool of FlowJo V.X. The data measured from unstained Hek293 cells (background fluorescence) was concatenated with data measured from cells that had been stained with a single fluorescent dye *in silico*. As a result, the data consist of two populations, for example a GFP$^+$ and GFP$^-$ population. Compensation is calculated so that the median fluorescence signal in the respective other parameters (here PE) does not differ for GFP$^+$ and GFP$^-$ cells (see Fig. E.2). The theory of fluorescence compensation is described in detail in the next section.

Single cell measurements have been mainly performed to analyse the correlation of absolute Erk expression values and Erk phosphorylation. However, as larger cells have increased levels of protein, it can be expected that different signals measured from a population of cells are always correlated. To assess the correlation of fluorescence parameters independent of morphological properties like size and granularity, the data have been transformed using a regression model as described [86].

E.2 The principle of fluorescence compensation

I follow the mathematical description of [85]. In presence of fluorescence spillover the observed signal O_{ij} of parameter j in an experiment i is the sum of the true signal T_{ij} and some percentage of the signals from the other parameters. The goal of compensation is to find out how great these percentages are and to correct the signal by subtracting these parts. For a two colour compensation the true and the observed signal in an experiment i have the following relation:

$$
\begin{aligned}
T_{i1} &= O_{i1} \cdot f_{11} + O_{i2} \cdot f_{21} \\
T_{i2} &= O_{i1} \cdot f_{12} + O_{i2} \cdot f_{22} \, .
\end{aligned}
\tag{E.1}
$$

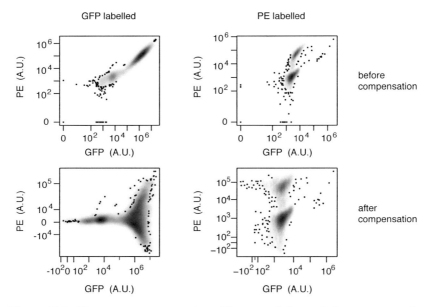

Figure E.2 Fluorescence compensation. The upper left panel shows a scatterplot of the GFP signal (FL1.A) and the PE signal (FL2.A) of GFP$^-$ cells (uncorrelated population) and GFP$^+$ cells (population with strong correlation of FL1.A and FL2.A) before compensation. After compensation, shown in the panel below, the median of both populations in the PE parameter is identical. The upper right panel shows the population of PE$^-$ cells (Hek293 WT) and Hek293 cells stained only with PE. After compensation, shown in the panel below, the median GFP signal is aligned. The data have been transformed using a biexponential function, to accommodate negative values that arise after compensation of the signal [119].

In this equation $-f_{kj}$ describes the fraction of parameter k that spills over to parameter j. The coefficients f_{kj} make up the so-called compensation matrix F, which can be seen when the two equations (E.1) are written in matrix notation:

$$\begin{pmatrix} T_{i1} & T_{i2} \end{pmatrix} = \begin{pmatrix} O_{i1} & O_{i2} \end{pmatrix} \cdot \begin{pmatrix} f_{1,1} & f_{1,2} \\ f_{2,1} & f_{2,2} \end{pmatrix} . \tag{E.2}$$

Solving this system is only possible, when measurements are available where the correlation between true and observed signal is known. This can be realised by measuring samples that contain only one stain. For example when the cells are only stained with GFP, the observed GFP signal is equal to the true GFP signal and the true signal of all other parameters has to be zero, in mathematical terms, $T_{ij} = O_{ij}$ if $i = j$, else $T_{ij} = 0$ (e.g. where experiment $i = 1$ refers to an experiment in which cells are only stained with a fluorophore that is measured in parameter $j = 1$). Of course, the true signal can only equal zero, when the background fluorescence has been subtracted beforehand. The equation system containing as many rows as single stain experiments,

$$\begin{pmatrix} O_{11} & 0 \\ 0 & O_{22} \end{pmatrix} = \begin{pmatrix} O_{11} & O_{12} \\ O_{21} & O_{22} \end{pmatrix} \cdot \begin{pmatrix} f_{1,1} & f_{1,2} \\ f_{2,1} & f_{2,2} \end{pmatrix} , \tag{E.3}$$

can now be solved for the compensation matrix. All observed signals O_{ij} in equation (E.3) are the median fluorescence values (corrected by the respective median background fluorescence) from the single-stain controls $i = 1, 2$. Fluorescence compensation is performed by multiplying the data matrix (where each column represents one fluorescence parameter and the rows correspond to the single measurements) with the compensation matrix.

E.3 Hek293 cells kept in 10% FCS have low levels of phosphoErk

To quantify Erk phosphorylation in single cells I have used an anti-ppErk1/2 antibody that is conjugated to the fluorophore Alexa647. An equally labelled antibody of the same isotype with no specific target (called isotype control) is expected to show signals above background due to unspecific binding. Specific binding will increase the signal even further. Hek293 cells, which are kept in presence of 10% FCS, have phosphoErk levels below the detection limit in Westernblots, the Bioplex assay and in isoelectric focusing experiments performed in this study. However, the signal of phosphoErk is significantly shifted to higher values compared to the distribution of the isotype control signal in the flow cytometry analysis (Fig. E.3). Most likely Erk signalling is not fully shut down in Hek293 cells in the presence of FCS, but the levels of Erk phosphorylation are very low. Targeted stimulation with EGF increases phosphoErk levels (see signal distribution at 5 min after stimulation in Fig. E.3), so they become visible in other less sensitive antibody-based assays.

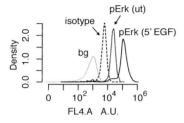

Figure E.3 Shown are the signal distributions in FL4.A for Hek293 cells without any staining (light grey), Hek293 cells labelled with anti-ppErk-Alexa647 when they were either untreated (grey) or stimulated with EGF for 5 min (black) and Hek293 cells stained with an isotype control (according to type of anti-ppErk antibody) conjugated to Alexa647 (dashed line).

F RT-PCR

RNA was isolated from cells 0.5, 1, 1.5 and 2 hours after treatment with 25 ng/ml EGF using the RNeasy Mini Kit (Qiagen) according to the suppliers' protocol. Concentration of RNA was determined on a Qubit Fluorometer using the Qubit RNA BR Assay Kit (Life Technologies). From each sample 1 μg RNA was reverse transcribed to cDNA using the Transcriptor High Fidelity cDNA Synthesis Kit (Roche). Quantitative real-time PCR analysis was performed using a StepOnePlus 96 well format Light-Cycler apparatus (Applied Biosystems). Control gene and target gene were amplified in one well using the TaqMan probe for PGK1 Hs00943174_g1 (VIC label) in combination with the probe for EGR1, Hs00152928_m1 (FAM label), or c-Fos, Hs00170630_m1 (FAM label). All samples were run in technical duplicates and three biological replicates.

G Quantification of impedance-based growth assays

G.1 Proliferation measurements using the xCELLigence technology

A comprehensive paper on the analysis of growth curves from impedance-based assays has been published in Bioinformatics [164]. This chapter constitutes a part of this publication. Measurement of proliferation of HT29 and RKO cells was performed by Raphaela Fritsche-Guenther, I have performed the measurements of Hek293 cells and evaluated the data.

The commercially available system xCELLigence [76] measures the proliferation of adherent cells in E-Plates. These plates are basically 96-well culture dishes with the special feature of integrated electrode structures at the bottom of each well. Via these electrodes, a small alternating potential is applied, and the impedance of the circuit is measured. The system calculates a so-called cell index, c_i, which is proportional to the impedance from which the background (i.e. impedance of medium without cells) is subtracted. This index increases when the electrode structures become covered with adherent cells. Thus, the cell index is not a direct function of cell number, but rather of the collective cell surface, the mode of cell attachment and cell membrane composition. Therefore it is important to establish that the increase in the cell index is primarily due to the formation of new adherent cells.

To determine if the ci faithfully reports cell numbers, we used 96-well plates in which the central 4 rows of the electrode structures in each well were left out, allowing to inspect relative confluency using microscopy. In these plates, we seeded either $5 \cdot 10^3$, 10^4 or $2 \cdot 10^4$ cells per well, which we either left untreated (triplicates per cell line, HT29 and RKO) or treated with Mek-inhibitor U0126 or

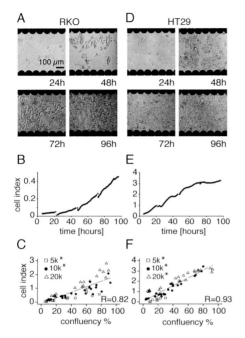

Figure G.1 Cell index correlates with cell surface and reflects proliferation. Analysis of correlation of cell index and % confluency of RKO cells in A) to C) and of HT29 in D) to F). Microscopic images show proliferation (A,D) which correlates with increasing cell index values in the corresponding growth curves in panel B) and E). The curves are broken at times when the plate was removed for microscopic analysis. The correlation of % confluency (as estimated by automated image segmentation) and measured cell index from 20 growth curves at 4 time points for each cell line, RKO and HT29, respectively is resolved by the number of seeded cells in panel C) and F) (5k cells = 5000 cells). The Pearson correlation of all data is indicated with R. A significantly improved linear correlation of a subgroup of the data is indicated with ∗ in the figure legends.

its vehicle control DMSO (both in duplicates per cell line for cells seeded at 10^4 or $2 \cdot 10^4$ cells). Growth curves were recorded over a time period of 4 days with daily interruptions for microscopic analysis to estimate % confluency. Figures G.1A and D show examples of bright field images taken from the electrode-free zone for untreated RKO ($5 \cdot 10^3$ cells/well initially) and HT29 ($2 \cdot 10^4$ cells/well initially) at the indicated time points. The corresponding growth curves are shown in Fig. G.1B and E.

To determine the relative confluency in an unbiased way, I used automated image segmentation to find the borders of cells. I implemented a routine using functions of the Image Processing Toolbox of MATLAB and the single steps of this algorithm are described in Fig. G.2. Briefly, the gold electrode structures, which appear as well defined dark objects in the pictures, were detected by k-means clustering of the luminosity layer and removed from the images thereafter (step 1 in Fig. G.2). Cells were detected using texture segmentation, i.e. by applying a range filter that amplifies intensity variations in the neighbourhood

Step 1 | Isolate the electrode-free area

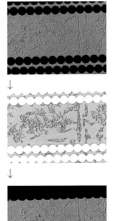

I have used *functions* of the image processing toolbox from MATLAB to perform image segmentation. In the first step, the image is converted from RGB to the Lab color space.

The luminosity layer is divided into 4 different levels of luminosity using kmeans clustering, as indicated by 4 different shades of grey in the image to the left. The electrode structures are recognised as the cluster with the lowest luminosity.

Next the image is converted to a binary image, by setting the lowest luminosity cluster to 1 and all other clusters to zero. The electrode-free area still contains some salt and pepper noise, which can be removed using the function *bwareaopen*. Finally, the binary image can be used to create a mask that accesses the electrode-free area, as shown on the left.

Step 2 | Isolate surface covered by cells

Texture segmentation was used to separate cells from the background. Each pixel has at maximum 8 neighbouring pixels. The *rangefilt* image transformation calculates the absolute difference of the maximal and minimal pixel value within the neighbourhood (the so-called range) and assigns this intensity range to the pixel. Image areas with only slight variations of pixel intensities, like the background, assume values close to zero, which appear as black in the image to the left.

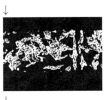

Using automatic thresholding (function *graythresh*) the image is converted to a binary image in which background and cellular structures are separated from each other.

However, there are still some disconnected contours, which can be closed by dilation of contour pixels. Then all closed cellular outlines are filled using the function *imfill*. Because of dilation the outline of cells is overestimated. By erosion of contours this effect can be partly reversed.

Figure G.2   Algorithm of automated image segmentation.

of an image pixel (step 2 in Fig. G.2). The relative confluency was determined by counting total image pixels and pixels that make up the segmented areas.

The relation of % confluency with the measured cell index is shown in Figures G.1C and F for RKO and HT29, respectively. There is a clear linear correlation for the cell line HT29 with a Pearson correlation coefficient of 0.93. For very high relative confluency (close to 100%) the curve shows saturation. The reason for this is unclear. Estimation errors may arise as % confluency of the whole well is extrapolated from the small proportion of the well that is accessible for microscopy. Also, in a previous study it was found that in general the cell index represents cell number only for sub-confluent cultures [144]. Consequently, impedance growth curves should not be evaluated as proportional to cell number after the curve has approached a plateau.

For RKO cells, there are stronger deviations from linearity, the Pearson correlation was 0.82 when all data points are taken into account. Again, deviation resulted mainly from data at high % confluency.

Visual inspection of the data shown in Fig. G.1C and F suggested that correlation within the groups of cells seeded at the same density is higher than overall correlation. The cell index seems to have shifted to higher values in a systematic fashion with increasing numbers of seeded cells (subgroups with significantly improved correlation are marked with a star in Fig. G.1 C and F, refer to [164] for details of the analysis of significance). One interpretation is that initial cell density influences the attachment mode and thus also subsequent measurements of the cell index. These results suggest that generally assays should be started with the same number of seeded cells to make them comparable.

It can be concluded that the cell index reports % confluency of RKO and HT29 cells with acceptable accuracy, as long as the culture is sub-confluent and the experimenter seeds the same amount of cells in each well. The % confluency also relates to cell number, as long as the cells are far from confluency where their contact surface is significantly decreased [164].

G.2 Model fits of Hek293 growth curves

In the main text I show one exemplary fit of Hek293 growth curves for each transfection type in Fig. 3.11. Here the fits for all 3 replicates per cell line are shown in Fig. G.3.

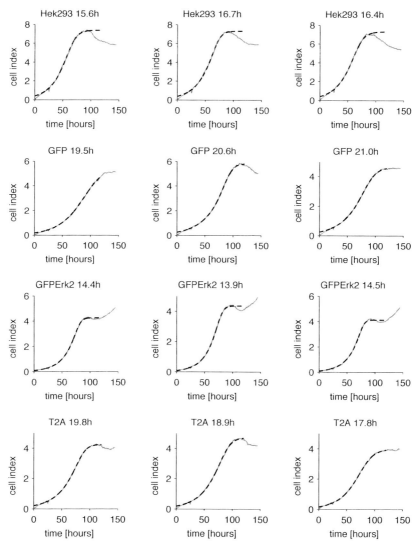

Figure G.3 WT Hek293 cells or Hek293 cells expressing GFP, Erk2 (labelled with T2A here) or GFPErk2 have been seeded at $5 \cdot 10^3$ cells/well and growth was followed in real time using the impedance-based technology xCELLigence. The Richards growth model (black dashed lines) was fitted to the measured growth curves (grey lines). The cell cycle time derived from the model parameters according to eq. (3.11) is indicated at the top of each panel.

H Additional material on modelling for chapter 2

H.1 From partial and total derivatives to response coefficients

Response coefficients are nothing but normalised derivatives, just like control coefficients that serve in the analysis of metabolic networks. The local response corresponds to a normalised partial derivative, the global response is the normalised total derivative. Equations (2.8) and (2.9) express that the steady levels of ppMek and ppErk are functions of each other, as ppMek creates ppErk and ppErk feeds back to the ppMek level. Additionally, ppErk is a function of the total available Erk. Looking at changes of the steady state ppErk level with regard to infinitesimal small changes of Erk_T, we can write the following total differential:

$$\frac{d\text{ppErk}}{d\text{Erk}_T} = \frac{\partial \text{ppErk}}{\partial \text{Erk}_T} + \frac{\partial \text{ppErk}}{\partial \text{ppMek}} \cdot \frac{d\text{ppMek}}{d\text{Erk}_T}. \tag{H.1}$$

To obtain the global response coefficient, we need to multiply the equation with Erk_T/ppErk,

$$
\begin{aligned}
\frac{\text{Erk}_T}{\text{ppErk}} \frac{d\text{ppErk}}{d\text{Erk}_T} &= \frac{\text{Erk}_T}{\text{ppErk}} \frac{\partial \text{ppErk}}{\partial \text{Erk}_T} + \frac{\text{Erk}_T}{\text{ppErk}} \frac{\partial \text{ppErk}}{\partial \text{ppMek}} \cdot \frac{d\text{ppMek}}{d\text{Erk}_T} \\
&= \frac{\text{Erk}_T}{\text{ppErk}} \frac{\partial \text{ppErk}}{\partial \text{Erk}_T} + \frac{\text{ppMek}}{\text{ppMek}} \frac{\text{Erk}_T}{\text{ppErk}} \frac{\partial \text{ppErk}}{\partial \text{ppMek}} \cdot \frac{d\text{ppMek}}{d\text{Erk}_T} \\
&= \frac{\text{Erk}_T}{\text{ppErk}} \frac{\partial \text{ppErk}}{\partial \text{Erk}_T} + \frac{\text{ppMek}}{\text{ppErk}} \frac{\partial \text{ppErk}}{\partial \text{ppMek}} \cdot \frac{\text{Erk}_T}{\text{ppMek}} \frac{d\text{ppMek}}{d\text{Erk}_T} \\
&= r_{\text{Erk}_T}^{\text{ppErk}} + r_{\text{ppMek}}^{\text{ppErk}} \cdot R_{\text{Erk}_T}^{\text{ppMek}}
\end{aligned}
$$

This equation corresponds to the model equation (2.10), which makes use of response coefficients.

H.2 Modelling inhibition of Mek

Small molecule Mek inhibitors like U0126 or AZD6244 bind non-competitively with ATP and thus reach a very high specificity for the kinase Mek, in contrast to other inhibitors which show a high cross-specificity for various kinases. It is assumed that these inhibitors bind Mek independently of its activation status, which means that the inhibitor basically reduces the amount of total available Mek. In consequence, a very simple model can be used to derive the amount of inhibitor-bound Mek depending on the dissociation constant K_D and the concentration of the inhibitor I_T.

$$\text{Mek} + \text{I} \underset{k_2}{\overset{k_1}{\rightleftharpoons}} \text{Mek}^{\text{I}} \quad \text{with} \quad K_D = \frac{k_2}{k_1} \tag{H.2}$$

The change of Mek^{I} with time has to equal zero in steady state:

$$
\begin{aligned}
\frac{\mathrm{d}\text{Mek}^{\text{I}}}{\mathrm{d}t} &= k_1 \cdot \text{Mek} \cdot \text{I} - k_2 \cdot \text{Mek}^{\text{I}} \\
&= k_1 \left(\text{Mek}_{\text{T}} - \text{Mek}^{\text{I}} \right) \left(\text{I}_{\text{T}} - \text{Mek}^{\text{I}} \right) - k_2 \cdot \text{Mek}^{\text{I}} \overset{!}{=} 0 \,.
\end{aligned}
$$

Mek^{I} can assume two different values,

$$\text{Mek}^{\text{I}}_{1/2} = \frac{\text{Mek}_{\text{T}} + \text{I}_{\text{T}} + K_D}{2} \pm \sqrt{\frac{\left(\text{Mek}_{\text{T}} + \text{I}_{\text{T}} + K_D \right)^2}{4} - \text{Mek}_{\text{T}} \cdot \text{I}_{\text{T}}} \,,$$

however, $\left(\text{Mek}_{\text{T}} + \text{I}_{\text{T}} \right) / 2$ represents the mean value of Mek_{T} and I_{T} and thus is always bigger than either Mek_{T} or I_{T}. Mek^{I} cannot exceed the level of total inhibitor or the total amount of Mek, and thus, only the solution with "-" is correct:

$$\text{Mek}^{\text{I}} = \frac{\text{Mek}_{\text{T}} + \text{I}_{\text{T}} + K_D}{2} - \sqrt{\frac{\left(\text{Mek}_{\text{T}} + \text{I}_{\text{T}} + K_D \right)^2}{4} - \text{Mek}_{\text{T}} \cdot \text{I}_{\text{T}}} \,. \tag{H.3}$$

The dissociation constant K_D is known to be in the nanomolar range, so an approximation of Mek^{I} can be obtained by assuming $K_D = 0$, which means that

$k_2 = 0$. With that assumption the differential equation simplifies to

$$
\begin{aligned}
\frac{\mathrm{dMek^I}}{\mathrm{d}t} &= k_1 \cdot \mathrm{Mek} \cdot \mathrm{I} - k_2 \cdot \mathrm{Mek^I} \\
&= k_1 \left(\mathrm{Mek_T} - \mathrm{Mek^I} \right) \left(\mathrm{I_T} - \mathrm{Mek^I} \right) \overset{!}{=} 0
\end{aligned}
$$

and obviously

$$
\mathrm{Mek^I} = \begin{cases} \mathrm{I_T} & \text{when } \mathrm{I_T} \leq \mathrm{Mek_T} \\ \mathrm{Mek_T} & \text{when } \mathrm{I_T} > \mathrm{Mek_T}. \end{cases}
$$

For the negative feedback model, I approximate $\mathrm{Mek^I} = \mathrm{I_T}$, which means that the fraction of total Mek which is inhibitor bound equals $\mathrm{I} = \mathrm{Mek^I}/\mathrm{Mek_T} = \mathrm{I_T}/\mathrm{Mek_T}$. In section 2.7 I refer only to the inhibition of active Mek, which is ppMek, so that $\mathrm{Mek_T}$ has to be replaced with ppMek in the description above.

I Additional material on modelling for chapter 3

I.1 The dual phosphorylation cycle model

I have formulated a model of distributive Erk dual (de)phosphorylation according to the reaction scheme in Fig. 3.1 in section 3.2. The model has 9 components when counting all modification states of Erk and the various enzyme-substrate complexes. There are three conserved quantities, the total amount of the kinase Mek K_T, the total amount of phosphatase P_T and the total amount of Erk, called Erk_T. The ode system reads

$$
\begin{aligned}
\frac{d}{dt} C_1 &= k_{on1} \cdot Erk \cdot K - (k_{off1} + k_{cat1}) \cdot C_1 \\
\frac{d}{dt} C_2 &= k_{on2} \cdot pErk \cdot K - (k_{off2} + k_{cat2}) \cdot C_2 \\
\frac{d}{dt} D_1 &= k_{onp1} \cdot pErk \cdot P - (k_{offp1} + k_{catp1}) \cdot D_1 \\
\frac{d}{dt} D_2 &= k_{onp2} \cdot ppErk \cdot P - (k_{offp2} + k_{catp2}) \cdot D_2 \\
\frac{d}{dt} pErk &= k_{cat1} \cdot C_1 - k_{on2} \cdot pErk \cdot K + k_{off2} \cdot C_2 + k_{catp2} \cdot D_2 \\
&\quad - k_{onp1} \cdot pErk \cdot P + k_{offp1} \cdot D_1 \\
\frac{d}{dt} ppErk &= k_{cat2} \cdot C_2 - k_{onp2} \cdot ppErk \cdot P + k_{offp2} \cdot D_2
\end{aligned}
$$

where the concentrations of Erk, K and P are given by the conservation relations

$$
\begin{aligned}
K &= K_T - C_1 - C_2 \\
P &= P_T - D_1 - D_2 \\
Erk &= Erk_T - pErk - ppErk - C_1 - C_2 - D_1 - D_2 \ .
\end{aligned}
$$

Kinetic parameters

Most of the parameters have been measured in HeLa cells [6] (see table I.1). All rates that describe the formation of an enzyme-substrate complex ($k_{\text{on(p)}1/2}$) have been assumed to be identical, the same was assumed for the dissociation rates of these complexes ($k_{\text{off(p)}1/2}$). The number in the index of parameters refers to the modification number (1st or 2nd phosphorylation), a "p" is added to describe a reaction involving the phosphatase.

I.2 Implicit derivation of the pErk limit in a single and dual phosphorylation cycle

Single phosphorylation cycle

Goldbeter and Koshland derived the analytical solution for the stationary level of the components in a single phosphorylation cycle [57]. In principle their expressions can be used to derive the maximum activation level pErk_{max} and the Erk concentration where we see half maximal activation, $\text{Erk}_{\text{T,50}}$. However, there is a faster way. The derivation of pErk_{max} was already shown in section 3.2. We now need to determine the total protein level where the contribution of pErk is

$$\text{pErk} = \frac{K_{\text{M,P}}}{2 \left(\frac{v_{\text{max,P}}}{v_{\text{max,K}}} - 1 \right)} \, . \tag{I.1}$$

Placing expression (I.1) into equation (I.2) which is the steady state condition of a single phosphorylation cycle,

$$v_{\text{K}} = \frac{v_{\text{max,K}}(\text{Erk}_{\text{T,50}} - \text{pErk})}{K_{\text{M,K}} + (\text{Erk}_{\text{T,50}} - \text{pErk})} = \frac{v_{\text{max,P}} \cdot \text{pErk}}{K_{\text{M,P}} + \text{pErk}} = v_{\text{P}}, \tag{I.2}$$

we arrive at a protein concentration with half maximum activation

$$\text{Erk}_{\text{T,50}} = \frac{K_{\text{M,P}} + 2K_{\text{M,K}} \frac{v_{\text{max,P}}}{v_{\text{max,K}}}}{2 \left(\frac{v_{\text{max,P}}}{v_{\text{max,K}}} - 1 \right)} \approx K_{\text{M,K}} \tag{I.3}$$

that corresponds to $K_{\text{M,K}}$ approximately at weak basal signalling ($v_{\text{max,P}} \gg v_{\text{max,K}}$) and when $K_{\text{M,P}}$ is lower than $K_{\text{M,K}}$ as described in [96].

parameter $[\mathrm{s}^{-1}\,\mu\mathrm{M}^{-1}]$	value	comment
$k_{\mathrm{on}1}$	0.18	measured in [6]
$k_{\mathrm{on}2}$	0.18	assumed
$k_{\mathrm{onp}1}$	0.18	assumed
$k_{\mathrm{onp}2}$	0.18	assumed
$k_{\mathrm{cat}1}/K_{\mathrm{M}1}$	$3.9{\cdot}10^{-2}$	measured in [6]
$k_{\mathrm{cat}2}/K_{\mathrm{M}2}$	$2.1{\cdot}10^{-2}$	measured in [6]
parameter $[\mathrm{s}^{-1}]$	value	comment
d_1	$6.7{\cdot}10^{-3}$	pY-Erk $\rightarrow$ Erk [6]
d_2	$4.0{\cdot}10^{-3}$	pTpY-Erk $\rightarrow$ pY-Erk [6]
$k_{\mathrm{off}1}$	0.27	measured in [6]
$k_{\mathrm{off}2}$	0.27	assumed
$k_{\mathrm{offp}1}$	0.27	assumed
$k_{\mathrm{offp}2}$	0.27	assumed
$k_{\mathrm{cat}1}$	$7.47{\cdot}10^{-2}$	calculated
$k_{\mathrm{cat}2}$	$3.57{\cdot}10^{-2}$	calculated
$k_{\mathrm{catp}1}$	$5.85{\cdot}10^{-2}$	calculated
$k_{\mathrm{catp}2}$	$3.15{\cdot}10^{-2}$	calculated
parameter $[\mu\mathrm{M}]$	value	comment
Mek total	1.2	measured in [6]
Erk total	0.74	measured in [6]

Table I.1   Table on parameters used in the dual phosphorylation model. All parameters that were measured in [6] refer to HeLa cells. The so-called calculated catalytic rates have been obtained from the equation $k_{\mathrm{cat}} = \frac{r\cdot k_{\mathrm{off}}}{k_{\mathrm{on}}-r}$ where r corresponds to a measured apparent rate $k_{\mathrm{cat}}/K_{\mathrm{M}}$.

Dual phosphorylation cycle

The steady state of the dual phosphorylation cycle model above (section I.1) has no analytical solution. However, for very high levels of Erk the Michaelis-Menten model is a suitable approximation which can be used to derive the limit of single phosphorylated Erk.

Markevich et al. [103] used the Michaelis-Menten description and account for the distribution of kinase and phosphatase in two different reactions by including competitive inhibition terms. Leaving out sequestration effects by binding of the phosphatase to unphosphorylated Erk we can modify the equations of [103] to yield the following expression for the speed of the 1st phosphorylation and dephosphorylation in steady state:

$$v_{K1} = \frac{v_{max,K} \cdot Erk}{Erk + K_{M,K1}\left(1 + \frac{pErk}{K_{M,K2}}\right)} = \frac{v_{max,P} \cdot pErk}{pErk + K_{M,P1}\left(1 + \frac{ppErk}{K_{M,P2}}\right)} = v_{P1} \quad (I.4)$$

Assuming that the kinase is saturated in cycle 1, $v_{K1} = v_{max,K}$ and $ppErk = 0$. Like that (I.4) simplifies to

$$v_{max,K} = \frac{v_{max,P} \cdot pErk_{max}}{pErk_{max} + K_{M,P1}} \quad \leftrightarrow$$

$$pErk_{max} = \frac{K_{M,P1}}{\frac{v_{max,P}}{v_{max,K}} - 1}, \quad (I.5)$$

which correponds to the first eq. in (3.2).

I.3 Analytical approximation for the steady state of a dual phosphorylation cycle

The model of a dual phosphorylation cycle (equations in section I.1) was simplified to allow for the calculation of an analytical steady state. Assuming that the phosphatases never saturate, dephosphorylation reactions can be described by mass action laws and the components P, D_1 and D_2 vanish. When the dephosphorylation rate constant in the first/second cycle is called d_1/d_2, the ode system reads:

$$\frac{\mathrm{d}}{\mathrm{d}t}C_1 = k_{\mathrm{on1}} \cdot \mathrm{Erk} \cdot \mathrm{K} - (k_{\mathrm{off1}} + k_{\mathrm{cat1}}) \cdot C_1 \tag{I.6}$$

$$\frac{\mathrm{d}}{\mathrm{d}t}C_2 = k_{\mathrm{on2}} \cdot \mathrm{pErk} \cdot \mathrm{K} - (k_{\mathrm{off2}} + k_{\mathrm{cat2}}) \cdot C_2$$

$$\frac{\mathrm{d}}{\mathrm{d}t}\mathrm{pErk} = k_{\mathrm{cat1}} \cdot C_1 - k_{\mathrm{on2}} \cdot \mathrm{pErk} \cdot \mathrm{K} + k_{\mathrm{off2}} \cdot C_2 - d_1 \cdot \mathrm{pErk} + d_2 \cdot \mathrm{ppErk}$$

$$\frac{\mathrm{d}}{\mathrm{d}t}\mathrm{ppErk} = k_{\mathrm{cat2}} \cdot C_2 - d_2 \cdot \mathrm{ppErk},$$

which corresponds to

$$\frac{\mathrm{ppErk}}{C_2} = \frac{k_{\mathrm{cat2}}}{d_2}$$

$$\frac{\mathrm{pErk}}{C_1} = \frac{k_{\mathrm{cat1}}}{d_1}$$

$$\frac{\mathrm{Erk}}{C_1} = \frac{K_{\mathrm{M1}}}{\mathrm{K}}$$

$$\frac{\mathrm{pErk}}{C_2} = \frac{K_{\mathrm{M2}}}{\mathrm{K}}$$

$$\mathrm{K} \approx \mathrm{K_T} - C_1 \tag{I.7}$$

$$\mathrm{Erk} \approx \mathrm{Erk_T} - C_1 - \mathrm{pErk} \tag{I.8}$$

in steady state. It is further assumed that the amounts of ppErk and C_2 are small, so that their contribution to $\mathrm{K_T}$ and $\mathrm{Erk_T}$ can be neglected, as was done in eq. (I.7) and eq. (I.8). This is the analytical solution:

$$C_1 = \frac{d_1(K_{\mathrm{M1}} + \mathrm{Erk_T}) + \mathrm{K_T}(d_1 + k_{\mathrm{cat1}})}{2(d_1 + k_{\mathrm{cat1}})} \tag{I.9}$$

$$- \frac{\sqrt{[d_1(K_{\mathrm{M1}} + \mathrm{Erk_T}) + \mathrm{K_T}(d_1 + k_{\mathrm{cat1}})]^2 - 4(d_1 + k_{\mathrm{cat1}})d_1\mathrm{Erk_T}\mathrm{K_T}}}{2(d_1 + k_{\mathrm{cat1}})}$$

$$\mathrm{pErk} = \frac{k_{\mathrm{cat1}}}{d_1} \cdot C_1 \tag{I.10}$$

$$\mathrm{Erk} = \frac{K_{\mathrm{M1}} \cdot C_1}{\mathrm{K_T} - C_1} \tag{I.11}$$

$$\mathrm{ppErk} = \frac{k_{\mathrm{cat1}}k_{\mathrm{cat2}} \cdot C_1(\mathrm{K_T} - C_1)}{d_1 d_2 K_{\mathrm{M2}}} \tag{I.12}$$

$$C_2 = \frac{k_{\mathrm{cat1}} \cdot C_1(\mathrm{K_T} - C_1)}{d_1 K_{\mathrm{M2}}} \tag{I.13}$$

$$\mathrm{K} = \mathrm{K_T} - C_1. \tag{I.14}$$

Now, I want to derive the maximal level of ppErk with respect to Erk_T. Equation (I.12) can be simplified by summarising all kinetic constants with γ:

$$
\begin{aligned}
\text{ppErk}(\text{Erk}_T) &= \frac{k_{\text{cat1}}k_{\text{cat2}} \cdot C_1(\text{Erk}_T)(K_T - C_1(\text{Erk}_T))}{d_1 d_2 K_{M2}} \\
&= \gamma \cdot C_1(\text{Erk}_T)(K_T - C_1(\text{Erk}_T)) \,.
\end{aligned}
$$

The derivative of ppErk with respect to Erk_T comes down to the derivative of C_1 with respect to Erk_T:

$$
\begin{aligned}
\frac{d}{d\text{Erk}_T} \text{ppErk}(\text{Erk}_T) &= \gamma \left[\frac{dC_1}{d\text{Erk}_T}[K_T - C_1(\text{Erk}_T)] - C_1 \frac{dC_1}{d\text{Erk}_T} \right] \\
&= \gamma \frac{dC_1}{d\text{Erk}_T}[K_T - 2C_1(\text{Erk}_T)]
\end{aligned}
$$

where

$$
\frac{d}{d\text{Erk}_T} \text{ppErk}(\text{Erk}_T) = 0 \leftrightarrow \frac{dC_1}{d\text{Erk}_T} = 0 \vee C_1 = \frac{K_T}{2} \,.
$$

As C_1 grows with the amount of Erk_T until saturation of the kinase with Erk, the first condition, $\frac{dC_1}{d\text{Erk}_T} = 0$, is never fulfilled. However, when

$$
C_1 = \frac{K_T}{2}
$$

then according to equation (I.8) the maximum level of ppErk is found at

$$
\text{Erk}_T = K_{M1} + \left(1 + \frac{k_{\text{cat1}}}{d_1} \right) \frac{K_T}{2}
$$

and the maximal level of ppErk itself is calculated from eq. (I.12) to

$$
\text{ppErk}_{\text{max}} = \frac{k_{\text{cat1}}k_{\text{cat2}} \cdot K_T^2}{4 d_1 d_2 K_{M2}} \,.
$$

In summary, the approximate maximum coordinate (eq. (3.7) in main text) reads

$$
(\text{Erk}_T, \text{ppErk})_{\text{max}} = \left(K_{M1} + \left[1 + \frac{k_{\text{cat1}}}{d_1} \right] \frac{K_T}{2} , \frac{k_{\text{cat1}}k_{\text{cat2}}}{d_1 d_2 K_{M2}} \cdot \frac{K_T^2}{4} \right) \,.
$$

In this model C_1 approaches the level of K_T for increasing concentrations of Erk. It follows from equation (I.10) that the maximal amount of singly phosphorylated Erk refers to:

$$\text{pErk}_{\max} = \frac{k_{\text{cat}1}}{d_1} \cdot \text{K}_\text{T} .\tag{I.15}$$

I.4 Mathematical models for the differential transient phosphoErk response in WT and Erk$^+$ cells

The model at the basis of the model variations in Fig. 3.23 and 3.25 from chapter 3.5 reads

$$
\begin{aligned}
\frac{\text{d}}{\text{d}t}\, \text{C}_1 &= k_{\text{on}} \cdot \text{Erk} \cdot \text{K}(t) - (k_{\text{off}} + k_{\text{cat}}) \cdot \text{C}_1 \\
\frac{\text{d}}{\text{d}t}\, \text{C}_2 &= k_{\text{on}} \cdot \text{pErk} \cdot \text{K}(t) - (k_{\text{off}} + k_{\text{cat}}) \cdot \text{C}_2 \\
\frac{\text{d}}{\text{d}t}\, \text{pErk} &= k_{\text{cat}} \cdot \text{C}_1 - k_{\text{on}} \cdot \text{pErk} \cdot \text{K}(t) + k_{\text{off}} \cdot \text{C}_2 - d \cdot \text{pErk} + d \cdot \text{ppErk} \\
\frac{\text{d}}{\text{d}t}\, \text{ppErk} &= k_{\text{cat}} \cdot \text{C}_2 - d \cdot \text{ppErk}
\end{aligned}
$$

with the conservation relations

$$
\begin{aligned}
\text{Erk} &= \text{Erk}_{\text{T,meas.}} - \text{C}_1 - \text{C}_2 - \text{pErk} - \text{ppErk} \tag{I.16} \\
\text{K}(t) &= \text{ppMek1}_{\text{meas.}}(t) - \text{C}_1 - \text{C}_2 . \tag{I.17}
\end{aligned}
$$

The total amount of Erk, $\text{Erk}_{\text{T,meas.}}$ in eq. (I.16) was quantified from Western blots (A.U.). The time dependent activity of Mek measured with the bioplex assay (A.U.), $\text{ppMek1}_{\text{meas.}}(t)$ in eq. (I.17) provides an inhomogeneous term in the ode system. In the following list I show how the basic model was modified to obtain the different models that were analysed.

model A in Fig. 3.23

- ppMek1 as measured from WT cells is used in eq. (I.17) to fit WT and Erk$^+$ data of ppErk model B in Fig. 3.23

- cell type-specific measured ppMek1 signals are used in eq. (I.17) for fitting of either WT or Erk^+ data of ppErk

model C in Fig. 3.23

- cell type-specific measured ppMek1 signals are used in eq. (I.17) for fitting of either WT or Erk^+ data of ppErk

- the rate constant d is replaced by $d_{\text{WT}}/d_{\text{Erk}^+}$ for the fit of WT data/Erk^+ data of ppErk

model in Fig. 3.25

- cell type-specific measured ppMek1 signals are used in eq. (I.17) for fitting of either WT or Erk^+ data of ppErk

- the dephosphorylation rate $d \cdot (\text{p})\text{pErk}$ is replaced by $\frac{v_{\text{max}} \cdot (\text{p})\text{pErk}}{K_{\text{M}} + (\text{p})\text{pErk}}$

model in Fig. 3.26

- the model fitted here corresponds to the ode system shown in section I.1 with $k_{\text{on1}} = k_{\text{on2}} = k_{\text{on}}$, $k_{\text{off1}} = k_{\text{off2}} = k_{\text{off}}$, $k_{\text{onp1}} = k_{\text{onp2}} = k_{\text{on,P}}$, $k_{\text{offp1}} = k_{\text{offp2}} = k_{\text{off,P}}$, $k_{\text{cat1}} = k_{\text{cat2}} = k_{\text{cat}}$, $k_{\text{catp1}} = k_{\text{catp2}} = k_{\text{cat,P}}$

- the conservation relations are replaced by $\text{Erk} = \text{Erk}_{\text{T,meas.}} - C_1 - C_2 - D_1 - D_2 - \text{pErk} - \text{ppErk}$, $\text{K}(t) = \text{ppMek1}_{\text{meas.}}(t) - C_1 - C_2$ and $\text{P} = \text{P}_{\text{T}} - D_1 - D_2$ where P_{T} is an additional parameter of the system

I.5 Parameter estimation

In this dissertation I evaluate the numerical solution of ode systems of the following form

$$\frac{\mathrm{d}\mathbf{x}(t, \mathbf{p})}{\mathrm{d}t} = \mathbf{f}\left(\mathbf{x}(t, \mathbf{p}), u(t), \mathbf{p}\right) \tag{I.18}$$

where $\mathbf{x}$ denotes a vector of N variables or molecular species, $\mathbf{p}$ summarises the P parameters of the system and $\mathbf{u}(t)$ represents inhomogeneities, e.g. a stimulus that activates a signalling system.

To minimise the deviation of the model from the experimental data I have used the MATLAB function lsqnonlin, which uses a trust-region-reflective algorithm. As a default the algorithm calculates the influence of a parameter on the objective function by a finite difference estimation - the ode system is solved numerically for two different parameter sets.

$$\frac{\mathrm{d}\mathbf{x}(t, \mathbf{p})}{\mathrm{d}p_i} = \frac{\mathbf{x}(t, \mathbf{p}) - \mathbf{x}(t, \mathbf{p} + h \cdot \mathbf{e_i})}{h} \tag{I.19}$$

In theory, the smaller h, the more accurate is the approximation in equation (I.19). However, for very small values of h the two ode system solutions will be almost identical, and the error inherent to numerical solutions becomes significant, as described in [126]. To avoid numerical instabilities or systematic errors in sensitivity calculation I have used FSA, forward sensitivity analysis [41]. Calculating the derivative of the ode system (I.18) with respect to $\mathbf{p}$ we obtain

$$\frac{\mathrm{d}}{\mathrm{d}t}\frac{\mathrm{d}\mathbf{x}}{\mathrm{d}\mathbf{p}} = \frac{\partial \mathbf{f}}{\partial \mathbf{x}} \cdot \frac{\mathrm{d}\mathbf{x}}{\mathrm{d}\mathbf{p}} + \frac{\partial \mathbf{f}}{\partial \mathbf{p}}. \tag{I.20}$$

This is a set of $N \times P$ linear odes which describe the development of the sensitivities $\frac{\mathrm{d}\mathbf{x}}{\mathrm{d}\mathbf{p}} = S_{ij}$ over time. I have used the C-based solver CVODES with the help of the SundialsTB interface for MATLAB to solve the ode system (I.18) simultaneously with the sensitivity ode system (I.20). The sensitivities of the system variables can be used to derive the sensitivities of the objective function and these are supplied to the optimiser lsqnonlin.

Initial parameter sets for optimisation have been generated by latin hypercube sampling on an interval of [-3,3] in logarithmic space for each parameter. Each optimisation was run for 100 different initial parameter sets.

I.6 Simplistic Erk activation model with feedbacks

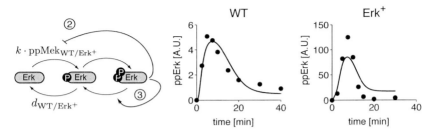

Figure I.1 Erk activity model with mass action laws for (de)phosphorylation. The cell type-specific measured level of ppMek1 times a rate constant k describes both phosphorylation reactions. Different rates of dephosphorylation were assumed for WT and Erk$^+$ cells (3 parameter model). Thus the model incorporates two mechanisms which are indicated by the encircled numbers according to description in Fig. 3.19. The panels on the right show the best fit (solid line) of the measured ppErk kinetics (dots) in WT and Erk$^+$ cells.

To prove the importance of ppMek sequestration for the successful fit of the model described in Fig. 3.23C I also tested a model which incorporates both feedbacks (mechanisms ② and ③ in Fig. I.1) but does not permit saturation of ppMek. Phosphorylation from Erk to pErk and from pErk to ppErk was assumed to proceed with a rate proportional to the level of ppMek1 as measured in WT and Erk$^+$ cells (see Fig. 3.22) multiplied with a rate constant k. This three parameter model cannot explain the data (see Fig. I.1).

I.7 Quasi steady state entails non-identifiable parameters

The model fits have been performed using the measured ppMek1 data as input function. The slow time scale on which ppMek1 peaks and goes back to basal levels is swiftly transmitted to the level of ppErk. Thus, for most of the time after stimulation, the ppErk levels are in a quasi steady state with respect to the ppMek levels. The result are pronounced parameter correlations, non-identifiabilities, respectively. The parameter optimisation was started from 100 different initial parameter sets (latin hypercube sampling between [-3,3] in logarithmic space for each parameter). The best 20 fits where usually characterised

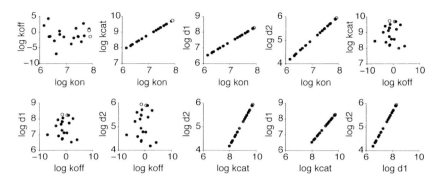

Figure I.2 Shown are all correlations of the parameters from the best 20 fits of the model in Fig. 3.23C. Parameter d_1 corresponds to d_{WT} and d_2 to parameter d_{Erk+}. The unfilled dots indicate the best two fits.

by objective function values that differed only at the 5th decimal place. Figure I.2 shows the parameter value correlations for the best 20 fits of the model from Fig. 3.23C. Due to the quasi steady state of ppErk, parameters that characterise ppErk formation (k_{on} and k_{cat}) form non-indentifiable pairs with parameters for deactivation d_1 and d_2. The dephosphorylation rate constant d_1 corresponds to the one found for WT cells, d_2 is the deactivation rate constant in Erk$^+$ cells. As each deactivation rate constant correlates e.g. with k_{cat}, d_1 also correlates with d_2. Parameter non-identifiabilities may decrease performance of the optimisation routine, however they do not invalidate the herein used method of hypothesis testing.

J Abbreviations

AKT	v-akt murine thymoma viral oncogene homolog, isoforms AKT1,2 and 3
c-Fos	FBJ murine osteosarcoma viral oncogene homolog, a transcription factor
c-Jun	v-jun avian sarcoma virus 17 oncogene homolog, a transcription factor
CK2α	casein kinase 2 alpha
DUSP	dual-specificity phosphatase
EGF	epidermal growth factor
EGFR	EGF receptor, also known as ErbB1 (erythroblastic leukemia viral oncogene homolog 1)
EGR1	early growth response 1, a transcription factor
Elk1	ETS-like gene 1, a transcription factor of the ETS (E26 transformation-specific) family
ErbB	see HER
Erk(2)$^+$ cells	Hek293 cells with Erk2 overexpression
FACS	fluorescence-activated cell sorting
FSC-A/H	forward scattered light signal, area/height of the pulse
Grb2	growth factor receptor-bound protein 2, an adaptor protein
HCT116	a colon cancer derived cell line
Hek293	human embryonic kidney derived cell line
HeLa	a cervical cancer derived cell line
HER	human epidermal growth factor receptor family consisting of HER1 (EGFR,ErbB1), HER2 (ErbB2), HER3 (ErbB3) and HER4 (ErbB4)
HT29	a colon cancer derived cell line
KSR	kinase suppressor of Ras, isoforms KSR1,2
Lim1215	a colon cancer derived cell line
MAPKK	MAPK kinase

MAP2K1	Mek1
MAP2K2	Mek2
Mek	both isoforms (Mek1/2) of the MAPKK that phosphorylates and activates Erk
MP1	Mek-partner-1, a scaffold protein
Myc	v-myc avian myelocytomatosis viral oncogene homolog, a transcription factor
NGF	nerve growth factor
NIH3T3	a fibroblast cell line from *Mus musculus*
ode	ordinary differential equation
PC12	a cell line derived from a pheochromocytoma of the rat adrenal medulla
phosphoMek	is used to describe the amount of Mek detected by ppMek1/2-specific antibodies
PI3K	phosphatidylinositol-4-phosphate 3 kinase
PP2A	protein phosphatase 2A
ppMek	active Mek with phosphorylation on Ser^{218} (Ser^{222}) and on Ser^{222} (Ser^{226}) in Mek1 (Mek2)
p90RSK	AGC kinase of the RSK family
p70S6K	AGC kinase of the RSK family
PTP	the family of protein tyrosine phosphatases
Raf	v-raf-1 murine leukemia viral oncogene, a MAPKK kinase, isoforms A-Raf, B-raf and c-Raf (RAF1)
Ras	rat sarcoma viral oncogene, a GTPase, isoforms NRas, KRas and HRas
RKO	a colon cancer derived cell line
RSK	ribosomal protein S6 kinase
RT-PCR	real time PCR (polymerase chain reaction)
SDS-PAGE	sodium dodecyl sulfate - polyacrylamide gel electrophoresis
Sef	similar expression to fgf genes, a scaffold protein
Sos	son of sevenless, a guanine nucleotide exchange factor, isoforms Sos1,2
SSC-A/H	side scattered light signal, area/height of the pulse
SW480	a colon cancer derived cell line
ut	untreated

WT   wild type

Bibliography

[1] J. G. Albeck, G. B. Mills, and J. S. Brugge. Frequency-modulated pulses of ERK activity transmit quantitative proliferation signals. *Mol Cell*, 49(2):249–261, Dec 2012.

[2] U. Alon, M. G. Surette, N. Barkai, and S. Leibler. Robustness in bacterial chemotaxis. *Nature*, 397(6715):168–171, Jan 1999.

[3] I. Amit, A. Citri, T. Shay, Y. Lu, M. Katz, F. Zhang, G. Tarcic, D. Siwak, J. Lahad, J. Jacob-Hirsch, N. Amariglio, N. Vaisman, E. Segal, G. Rechavi, U. Alon, G. B. Mills, E. Domany, and Y. Yarden. A module of negative feedback regulators defines growth factor signaling. *Nat Genet*, 39(4):503–512, Apr 2007.

[4] N. G. Anderson, J. L. Maller, N. K. Tonks, and T. W. Sturgill. Requirement for integration of signals from two distinct phosphorylation pathways for activation of MAP kinase. *Nature*, 343(6259):651–653, Feb 1990.

[5] K. Aoki, K. Takahashi, K. Kaizu, and M. Matsuda. A quantitative model of ERK MAP kinase phosphorylation in crowded media. *Sci Rep*, 3:1541, Mar 2013.

[6] K. Aoki, M. Yamada, K. Kunida, S. Yasuda, and M. Matsuda. Processive phosphorylation of ERK MAP kinase in mammalian cells. *Proc Natl Acad Sci U S A*, 108(31):12675–12680, Aug 2011.

[7] R. Arvind, H. Shimamoto, F. Momose, T. Amagasa, K. Omura, and N. Tsuchida. A mutation in the common docking domain of ERK2 in a human cancer cell line, which was associated with its constitutive phosphorylation. *Int J Oncol*, 27(6):1499–1504, Dec 2005.

[8] F. A. Atienzar, K. Tilmant, H. H. Gerets, G. Toussaint, S. Speeckaert, E. Hanon, O. Depelchin, and S. Dhalluin. The use of real-time cell analyzer

technology in drug discovery: defining optimal cell culture conditions and assay reproducibility with different adherent cellular models. *J Biomol Screen*, 16(6):575–587, Jul 2011.

[9] R. Avraham and Y. Yarden. Feedback regulation of EGFR signalling: decision making by early and delayed loops. *Nat Rev Mol Cell Biol*, 12(2):104–117, Feb 2011.

[10] S. C. Baca, D. Prandi, M. S. Lawrence, J. M. Mosquera, A. Romanel, Y. Drier, K. Park, N. Kitabayashi, T. Y. MacDonald, M. Ghandi, E. Van Allen, G. V. Kryukov, A. Sboner, J.-P. Theurillat, T. D. Soong, E. Nickerson, D. Auclair, A. Tewari, H. Beltran, R. C. Onofrio, G. Boysen, C. Guiducci, C. E. Barbieri, K. Cibulskis, A. Sivachenko, S. L. Carter, G. Saksena, D. Voet, A. H. Ramos, W. Winckler, M. Cipicchio, K. Ardlie, P. W. Kantoff, M. F. Berger, S. B. Gabriel, T. R. Golub, M. Meyerson, E. S. Lander, O. Elemento, G. Getz, F. Demichelis, M. A. Rubin, and L. A. Garraway. Punctuated evolution of prostate cancer genomes. *Cell*, 153(3):666–677, Apr 2013.

[11] D. Bae and S. Ceryak. Raf-independent, PP2A-dependent MEK activation in response to ERK silencing. *Biochem Biophys Res Commun*, 385(4):523–527, Aug 2009.

[12] Z. Baharians and A. H. Schönthal. Autoregulation of protein phosphatase type 2A expression. *J Biol Chem*, 273(30):19019–19024, Jul 1998.

[13] A. Bansal, R. D. Ramirez, and J. D. Minna. Mutation analysis of the coding sequences of MEK-1 and MEK-2 genes in human lung cancer cell lines. *Oncogene*, 14(10):1231–1234, Mar 1997.

[14] H. Bendfeldt, M. Benary, T. Scheel, S. Frischbutter, A. Abajyan, A. Radbruch, H. Herzel, and R. Baumgrass. Stable IL-2 decision making by endogenous c-Fos amounts in peripheral memory T-helper cells. *J Biol Chem*, 287(22):18386–18397, May 2012.

[15] U. S. Bhalla. Signaling in small subcellular volumes. I. stochastic and diffusion effects on individual pathways. *Biophys J*, 87(2):733–744, Aug 2004.

[16] R. Bhargava, W. L. Gerald, A. R. Li, Q. Pan, P. Lal, M. Ladanyi, and B. Chen. EGFR gene amplification in breast cancer: correlation with epidermal growth factor receptor mRNA and protein expression and HER-2 status and absence of EGFR-activating mutations. *Mod Pathol*, 18(8):1027–1033, Aug 2005.

[17] N. Blüthgen. Transcriptional feedbacks in mammalian signal transduction pathways facilitate rapid and reliable protein induction. *Mol Biosyst*, 6(7):1277–1284, Jul 2010.

[18] N. Blüthgen, F. J. Bruggeman, S. Legewie, H. Herzel, H. V. Westerhoff, and B. N. Kholodenko. Effects of sequestration on signal transduction cascades. *FEBS J*, 273(5):895–906, Mar 2006.

[19] N. Blüthgen and S. Legewie. Systems analysis of MAPK signal transduction. *Essays Biochem*, 45:95–107, 2008.

[20] C. W. Brennan, R. G. W. Verhaak, A. McKenna, B. Campos, H. Noushmehr, S. R. Salama, S. Zheng, D. Chakravarty, J. Z. Sanborn, S. H. Berman, R. Beroukhim, B. Bernard, C.-J. Wu, G. Genovese, I. Shmulevich, J. Barnholtz-Sloan, L. Zou, R. Vegesna, S. A. Shukla, G. Ciriello, W. K. Yung, W. Zhang, C. Sougnez, T. Mikkelsen, K. Aldape, D. D. Bigner, E. G. Van Meir, M. Prados, A. Sloan, K. L. Black, J. Eschbacher, G. Finocchiaro, W. Friedman, D. W. Andrews, A. Guha, M. Iacocca, B. P. O'Neill, G. Foltz, J. Myers, D. J. Weisenberger, R. Penny, R. Kucherlapati, C. M. Perou, D. N. Hayes, R. Gibbs, M. Marra, G. B. Mills, E. Lander, P. Spellman, R. Wilson, C. Sander, J. Weinstein, M. Meyerson, S. Gabriel, P. W. Laird, D. Haussler, G. Getz, L. Chin, and T. C. G. A. R. N. . The somatic genomic landscape of glioblastoma. *Cell*, 155(2):462–477, Oct 2013.

[21] D. F. Brennan, A. C. Dar, N. T. Hertz, W. C. H. Chao, A. L. Burlingame, K. M. Shokat, and D. Barford. A Raf-induced allosteric transition of KSR stimulates phosphorylation of MEK. *Nature*, 472(7343):366–369, Apr 2011.

[22] J. M. Brondello, J. Pouysségur, and F. R. McKenzie. Reduced MAP kinase phosphatase-1 degradation after p42/p44mapk-dependent phosphorylation. *Science*, 286(5449):2514–2517, Dec 1999.

[23] M. C. Brown and C. E. Turner. Paxillin: adapting to change. *Physiol Rev*, 84(4):1315–1339, Oct 2004.

[24] F. J. Bruggeman, N. Blüthgen, and H. V. Westerhoff. Noise management by molecular networks. *PLoS Comput Biol*, 5(9):e1000506, Sep 2009.

[25] T. Brummer, P. Martin, S. Herzog, Y. Misawa, R. J. Daly, and M. Reth. Functional analysis of the regulatory requirements of B-Raf and the B-Raf(V600E) oncoprotein. *Oncogene*, 25(47):6262–6276, Oct 2006.

[26] T. Brummer, H. Naegele, M. Reth, and Y. Misawa. Identification of novel ERK-mediated feedback phosphorylation sites at the C-terminus of B-Raf. *Oncogene*, 22(55):8823–8834, Dec 2003.

[27] A. Brunet, G. Pagès, and J. Pouysségur. Constitutively active mutants of MAP kinase kinase (mek1) induce growth factor-relaxation and oncogenicity when expressed in fibroblasts. *Oncogene*, 9(11):3379–3387, Nov 1994.

[28] A. Brunet, G. Pagès, and J. Pouysségur. Growth factor-stimulated MAP kinase induces rapid retrophosphorylation and inhibition of MAP kinase kinase (MEK1). *FEBS Lett*, 346(2-3):299–303, Jun 1994.

[29] W. R. Burack and A. S. Shaw. Live cell imaging of ERK and MEK: simple binding equilibrium explains the regulated nucleocytoplasmic distribution of ERK. *J Biol Chem*, 280(5):3832–3837, Feb 2005.

[30] W. R. Burack and T. W. Sturgill. The activating dual phosphorylation of MAPK by MEK is nonprocessive. *Biochemistry*, 36(20):5929–5933, May 1997.

[31] S. Cagnol, E. Van Obberghen-Schilling, and J.-C. Chambard. Prolonged activation of ERK1,2 induces FADD-independent caspase 8 activation and cell death. *Apoptosis*, 11(3):337–346, Mar 2006.

[32] S. M. Carlson, C. R. Chouinard, A. Labadorf, C. J. Lam, K. Schmelzle, E. Fraenkel, and F. M. White. Large-scale discovery of ERK2 substrates identifies ERK-mediated transcriptional regulation by ETV3. *Sci Signal*, 4(196):rs11, Oct 2011.

[33] F. Catalanotti, G. Reyes, V. Jesenberger, G. Galabova-Kovacs, R. de Matos Simoes, O. Carugo, and M. Baccarini. A Mek1-Mek2 heterodimer determines the strength and duration of the Erk signal. *Nat Struct Mol Biol*, 16(3):294–303, Mar 2009.

[34] E. Cerami, J. Gao, U. Dogrusoz, B. E. Gross, S. O. Sumer, B. A. Aksoy, A. Jacobsen, C. J. Byrne, M. L. Heuer, E. Larsson, Y. Antipin, B. Reva, A. P. Goldberg, C. Sander, and N. Schultz. The cBio cancer genomics portal: an open platform for exploring multidimensional cancer genomics data. *Cancer Discov*, 2(5):401–404, May 2012.

[35] C. Cohen-Saidon, A. A. Cohen, A. Sigal, Y. Liron, and U. Alon. Dynamics and variability of ERK2 response to EGF in individual living cells. *Mol Cell*, 36(5):885–893, Dec 2009.

[36] M. Courcelles, C. Frémin, L. Voisin, S. Lemieux, S. Meloche, and P. Thibault. Phosphoproteome dynamics reveal novel ERK1/2 MAP kinase substrates with broad spectrum of functions. *Mol Syst Biol*, 9:669, May 2013.

[37] M. Cully, H. You, A. J. Levine, and T. W. Mak. Beyond PTEN mutations: the PI3K pathway as an integrator of multiple inputs during tumorigenesis. *Nat Rev Cancer*, 6(3):184–192, Mar 2006.

[38] P. B. Dennis, N. Pullen, R. B. Pearson, S. C. Kozma, and G. Thomas. Phosphorylation sites in the autoinhibitory domain participate in p70(s6k) activation loop phosphorylation. *J Biol Chem*, 273(24):14845–14852, Jun 1998.

[39] N. Dey, B. Leyland-Jones, and P. De. MYC-xing it up with PIK3CA mutation and resistance to PI3K inhibitors: summit of two giants in breast cancers. *Am J Cancer Res*, 5(1):1–19, Dec 2015.

[40] A. S. Dhillon, A. von Kriegsheim, J. Grindlay, and W. Kolch. Phosphatase and feedback regulation of Raf-1 signaling. *Cell Cycle*, 6(1):3–7, Jan 2007.

[41] G. R. J. Dickinson Robert P. Sensitivity analysis of ordinary differential equation systems - a direct method. *Journal of Computational Physics*, 21(2):123–143, Jun 1976.

[42] M. K. Dougherty, J. Müller, D. A. Ritt, M. Zhou, X. Z. Zhou, T. D. Copeland, T. P. Conrads, T. D. Veenstra, K. P. Lu, and D. K. Morrison. Regulation of Raf-1 by direct feedback phosphorylation. *Mol Cell*, 17(2):215–224, Jan 2005.

[43] S. T. Eblen, A. D. Catling, M. C. Assanah, and M. J. Weber. Biochemical and biological functions of the N-terminal, noncatalytic domain of extracellular signal-regulated kinase 2. *Mol Cell Biol*, 21(1):249–259, Jan 2001.

[44] B. Efron. Better bootstrap confidence intervals. *Journal of the American Statistical Association*, 82(No. 397):171–185, Mar. 1987.

[45] P. Eirew, A. Steif, J. Khattra, G. Ha, D. Yap, H. Farahani, K. Gelmon, S. Chia, C. Mar, A. Wan, E. Laks, J. Biele, K. Shumansky, J. Rosner, A. McPherson, C. Nielsen, A. J. L. Roth, C. Lefebvre, A. Bashashati, C. de Souza, C. Siu, R. Aniba, J. Brimhall, A. Oloumi, T. Osako, A. Bruna, J. L. Sandoval, T. Algara, W. Greenwood, K. Leung, H. Cheng, H. Xue, Y. Wang, D. Lin, A. J. Mungall, R. Moore, Y. Zhao, J. Lorette, L. Nguyen, D. Huntsman, C. J. Eaves, C. Hansen, M. A. Marra, C. Caldas, S. P. Shah, and S. Aparicio. Dynamics of genomic clones in breast cancer patient xenografts at single-cell resolution. *Nature*, 518(7539):422–426, Feb 2015.

[46] J. Elf and M. Ehrenberg. Fast evaluation of fluctuations in biochemical networks with the linear noise approximation. *Genome Res*, 13(11):2475–2484, Nov 2003.

[47] V. Emuss, M. Garnett, C. Mason, and R. Marais. Mutations of C-RAF are rare in human cancer because C-RAF has a low basal kinase activity compared with B-RAF. *Cancer Res*, 65(21):9719–9726, Nov 2005.

[48] J. E. Ferrell and R. R. Bhatt. Mechanistic studies of the dual phosphorylation of mitogen-activated protein kinase. *J Biol Chem*, 272(30):19008–19016, Jul 1997.

[49] C. Frémin and S. Meloche. From basic research to clinical development of MEK1/2 inhibitors for cancer therapy. *J Hematol Oncol*, 3:8, Feb 2010.

[50] B. B. Friday, C. Yu, G. K. Dy, P. D. Smith, L. Wang, S. N. Thibodeau, and A. A. Adjei. BRAF V600E disrupts AZD6244-induced abrogation of

negative feedback pathways between extracellular signal-regulated kinase and Raf proteins. *Cancer Res*, 68(15):6145–6153, Aug 2008.

[51] R. Fritsche-Guenther, F. Witzel, A. Sieber, R. Herr, N. Schmidt, S. Braun, T. Brummer, C. Sers, and N. Blüthgen. Strong negative feedback from Erk to Raf confers robustness to MAPK signalling. *Mol Syst Biol*, 7:489, May 2011.

[52] A. Fujioka, K. Terai, R. E. Itoh, K. Aoki, T. Nakamura, S. Kuroda, E. Nishida, and M. Matsuda. Dynamics of the Ras/ERK MAPK cascade as monitored by fluorescent probes. *J Biol Chem*, 281(13):8917–8926, Mar 2006.

[53] J. Gao, B. A. Aksoy, U. Dogrusoz, G. Dresdner, B. Gross, S. O. Sumer, Y. Sun, A. Jacobsen, R. Sinha, E. Larsson, E. Cerami, C. Sander, and N. Schultz. Integrative analysis of complex cancer genomics and clinical profiles using the cBioPortal. *Sci Signal*, 6(269):pl1, Apr 2013.

[54] T. P. J. Garrett, N. M. McKern, M. Lou, T. C. Elleman, T. E. Adams, G. O. Lovrecz, M. Kofler, R. N. Jorissen, E. C. Nice, A. W. Burgess, and C. W. Ward. The crystal structure of a truncated ErbB2 ectodomain reveals an active conformation, poised to interact with other ErbB receptors. *Mol Cell*, 11(2):495–505, Feb 2003.

[55] C. Ghiglione, N. Perrimon, and L. A. Perkins. Quantitative variations in the level of MAPK activity control patterning of the embryonic termini in Drosophila. *Dev Biol*, 205(1):181–193, Jan 1999.

[56] H. Gille, A. D. Sharrocks, and P. E. Shaw. Phosphorylation of transcription factor p62TCF by MAP kinase stimulates ternary complex formation at c-fos promoter. *Nature*, 358(6385):414–417, Jul 1992.

[57] A. Goldbeter and D. E. Koshland. An amplified sensitivity arising from covalent modification in biological systems. *Proc Natl Acad Sci U S A*, 78(11):6840–6844, Nov 1981.

[58] N. Gómez and P. Cohen. Dissection of the protein kinase cascade by which nerve growth factor activates MAP kinases. *Nature*, 353(6340):170–173, Sep 1991.

[59] F. A. Gonzalez, D. L. Raden, and R. J. Davis. Identification of substrate recognition determinants for human ERK1 and ERK2 protein kinases. *J Biol Chem*, 266(33):22159–22163, Nov 1991.

[60] F. L. Graham, J. Smiley, W. C. Russell, and R. Nairn. Characteristics of a human cell line transformed by DNA from human adenovirus type 5. *J Gen Virol*, 36(1):59–74, Jul 1977.

[61] K. M. Haigis, K. R. Kendall, Y. Wang, A. Cheung, M. C. Haigis, J. N. Glickman, M. Niwa-Kawakita, A. Sweet-Cordero, J. Sebolt-Leopold, K. M. Shannon, J. Settleman, M. Giovannini, and T. Jacks. Differential effects of oncogenic K-Ras and N-Ras on proliferation, differentiation and tumor progression in the colon. *Nat Genet*, 40(5):600–608, May 2008.

[62] D. Hanahan and R. A. Weinberg. The hallmarks of cancer. *Cell*, 100(1):57–70, Jan 2000.

[63] T. A. Haystead, P. Dent, J. Wu, C. M. Haystead, and T. W. Sturgill. Ordered phosphorylation of p42mapk by MAP kinase kinase. *FEBS Lett*, 306(1):17–22, Jul 1992.

[64] R. Heinrich, B. G. Neel, and T. A. Rapoport. Mathematical models of protein kinase signal transduction. *Mol Cell*, 9(5):957–970, May 2002.

[65] M. Hekman, A. Fischer, L. P. Wennogle, Y. K. Wang, S. L. Campbell, and U. R. Rapp. Novel C-Raf phosphorylation sites: serine 296 and 301 participate in Raf regulation. *FEBS Lett*, 579(2):464–468, Jan 2005.

[66] J. H. Her, S. Lakhani, K. Zu, J. Vila, P. Dent, T. W. Sturgill, and M. J. Weber. Dual phosphorylation and autophosphorylation in mitogen-activated protein (MAP) kinase activation. *Biochem J*, 296 (Pt 1):25–31, Nov 1993.

[67] J. K. Hériché, F. Lebrin, T. Rabilloud, D. Leroy, E. M. Chambaz, and Y. Goldberg. Regulation of protein phosphatase 2A by direct interaction with casein kinase 2alpha. *Science*, 276(5314):952–955, May 1997.

[68] L. A. Herzenberg, J. Tung, W. A. Moore, L. A. Herzenberg, and D. R. Parks. Interpreting flow cytometry data: a guide for the perplexed. *Nat Immunol*, 7(7):681–685, Jul 2006.

[69] F. R. Hirsch, M. Varella-Garcia, P. A. Bunn, M. V. D. Maria, R. Veve, R. M. Bremmes, A. E. Barón, C. Zeng, and W. A. Franklin. Epidermal growth factor receptor in non-small-cell lung carcinomas: correlation between gene copy number and protein expression and impact on prognosis. *J Clin Oncol*, 21(20):3798–3807, Oct 2003.

[70] G. Hornung and N. Barkai. Noise propagation and signaling sensitivity in biological networks: a role for positive feedback. *PLoS Comput Biol*, 4(1):e8, Jan 2008.

[71] J. Hu, E. C. Stites, H. Yu, E. A. Germino, H. S. Meharena, P. J. S. Stork, A. P. Kornev, S. S. Taylor, and A. S. Shaw. Allosteric activation of functionally asymmetric RAF kinase dimers. *Cell*, 154(5):1036–1046, Aug 2013.

[72] T. Hunter. Protein kinases and phosphatases: the yin and yang of protein phosphorylation and signaling. *Cell*, 80(2):225–236, Jan 1995.

[73] N. T. Ingolia. Topology and robustness in the Drosophila segment polarity network. *PLoS Biol*, 2(6):e123, Jun 2004.

[74] K. L. Jeffrey, M. Camps, C. Rommel, and C. R. Mackay. Targeting dual-specificity phosphatases: manipulating MAP kinase signalling and immune responses. *Nat Rev Drug Discov*, 6(5):391–403, May 2007.

[75] F. A. Karreth, G. M. DeNicola, S. P. Winter, and D. A. Tuveson. C-Raf inhibits MAPK activation and transformation by B-Raf(V600E). *Mol Cell*, 36(3):477–486, Nov 2009.

[76] N. Ke, X. Wang, X. Xu, and Y. A. Abassi. The xCELLigence system for real-time and label-free monitoring of cell viability. *Methods Mol Biol*, 740:33–43, Mar 2011.

[77] S. M. Keyse. Protein phosphatases and the regulation of mitogen-activated protein kinase signalling. *Curr Opin Cell Biol*, 12(2):186–192, Apr 2000.

[78] B. N. Kholodenko, J. B. Hoek, H. V. Westerhoff, and G. C. Brown. Quantification of information transfer via cellular signal transduction pathways. *FEBS Lett*, 414(2):430–434, Sep 1997.

[79] B. N. Kholodenko, A. Kiyatkin, F. J. Bruggeman, E. Sontag, H. V. Westerhoff, and J. B. Hoek. Untangling the wires: a strategy to trace functional
interactions in signaling and gene networks. *Proc Natl Acad Sci U S A*,
99(20):12841–12846, Oct 2002.

[80] C. Kiel and L. Serrano. Cell type-specific importance of ras-c-raf complex
association rate constants for MAPK signaling. *Sci Signal*, 2(81):ra38, Jul
2009.

[81] J. H. Kim, S.-R. Lee, L.-H. Li, H.-J. Park, J.-H. Park, K. Y. Lee, M.-K.
Kim, B. A. Shin, and S.-Y. Choi. High cleavage efficiency of a 2A peptide
derived from porcine teschovirus-1 in human cell lines, zebrafish and mice.
PLoS One, 6(4):e18556, Apr 2011.

[82] E. Kinoshita, E. Kinoshita-Kikuta, K. Takiyama, and T. Koike.
Phosphate-binding tag, a new tool to visualize phosphorylated proteins.
Mol Cell Proteomics, 5(4):749–757, Apr 2006.

[83] H. Kitano. Biological robustness. *Nat Rev Genet*, 5(11):826–837, Nov
2004.

[84] B. Klinger, A. Sieber, R. Fritsche-Guenther, F. Witzel, L. Berry, D. Schu-
macher, Y. Yan, P. Durek, M. Merchant, R. Schäfer, C. Sers, and N. Blüthgen. Network quantification of EGFR signaling unveils potential for targeted combination therapy. *Mol Syst Biol*, 9:673, Jan 2013.

[85] D. J. Klinke 2nd and K. M. Brundage. Scalable analysis of flow cytometry
data using R/Bioconductor. *Cytometry A*, 75(8):699–706, Aug 2009.

[86] T. A. Knijnenburg, O. Roda, Y. Wan, G. P. Nolan, J. D. Aitchison, and
I. Shmulevich. A regression model approach to enable cell morphology
correction in high-throughput flow cytometry. *Mol Syst Biol*, 7:531, Sep
2011.

[87] W. Kolch. Meaningful relationships: the regulation of the
Ras/Raf/MEK/ERK pathway by protein interactions. *Biochem J*, 351
Pt 2:289–305, Oct 2000.

[88] H. Kosako, N. Yamaguchi, C. Aranami, M. Ushiyama, S. Kose,
N. Imamoto, H. Taniguchi, E. Nishida, and S. Hattori. Phosphoproteomics

reveals new ERK MAP kinase targets and links ERK to nucleoporin-mediated nuclear transport. *Nat Struct Mol Biol*, 16(10):1026–1035, Oct 2009.

[89] W. J. Langlois, T. Sasaoka, A. R. Saltiel, and J. M. Olefsky. Negative feedback regulation and desensitization of insulin- and epidermal growth factor-stimulated p21ras activation. *J Biol Chem*, 270(43):25320–25323, Oct 1995.

[90] R. Leboeuf, J. E. Baumgartner, M. Benezra, R. Malaguarnera, D. Solit, C. A. Pratilas, N. Rosen, J. A. Knauf, and J. A. Fagin. BRAFV600E mutation is associated with preferential sensitivity to mitogen-activated protein kinase kinase inhibition in thyroid cancer cell lines. *J Clin Endocrinol Metab*, 93(6):2194–2201, Jun 2008.

[91] F. Lebrin, L. Bianchini, T. Rabilloud, E. M. Chambaz, and Y. Goldberg. CK2alpha-protein phosphatase 2A molecular complex: possible interaction with the MAP kinase pathway. *Mol Cell Biochem*, 191(1-2):207–212, Jan 1999.

[92] R. Lefloch, J. Pouysségur, and P. Lenormand. Single and combined silencing of ERK1 and ERK2 reveals their positive contribution to growth signaling depending on their expression levels. *Mol Cell Biol*, 28(1):511–527, Jan 2008.

[93] R. Lefloch, J. Pouysségur, and P. Lenormand. Total ERK1/2 activity regulates cell proliferation. *Cell Cycle*, 8(5):705–711, Mar 2009.

[94] S. Legewie, H. Herzel, H. V. Westerhoff, and N. Blüthgen. Recurrent design patterns in the feedback regulation of the mammalian signalling network. *Mol Syst Biol*, 4(1):190, Jan 2008.

[95] S. Legewie, B. Schoeberl, N. Blüthgen, and H. Herzel. Competing docking interactions can bring about bistability in the MAPK cascade. *Biophys J*, 93(7):2279–2288, Oct 2007.

[96] S. Legewie, C. Sers, and H. Herzel. Kinetic mechanisms for overexpression insensitivity and oncogene cooperation. *FEBS Lett*, 583(1):93–96, Jan 2009.

[97] J. A. Lehman and J. Gomez-Cambronero. Molecular crosstalk between p70S6k and MAPK cell signaling pathways. *Biochem Biophys Res Commun*, 293(1):463–469, Apr 2002.

[98] A. Levchenko, J. Bruck, and P. W. Sternberg. Scaffold proteins may biphasically affect the levels of mitogen-activated protein kinase signaling and reduce its threshold properties. *Proc Natl Acad Sci U S A*, 97(11):5818–5823, May 2000.

[99] J. W. Little, D. P. Shepley, and D. W. Wert. Robustness of a gene regulatory circuit. *EMBO J*, 18(15):4299–4307, Aug 1999.

[100] M. Mahalingam, R. Arvind, H. Ida, A. K. Murugan, M. Yamaguchi, and N. Tsuchida. ERK2 CD domain mutation from a human cancer cell line enhanced anchorage-independent cell growth and abnormality in drosophila. *Oncol Rep*, 20(4):957–962, Oct 2008.

[101] S. J. Mansour, W. T. Matten, A. S. Hermann, J. M. Candia, S. Rong, K. Fukasawa, G. F. V. Woude, and N. G. Ahn. Transformation of mammalian cells by constitutively active MAP kinase kinase. *Science*, 265(5174):966–970, Aug 1994.

[102] S. Marchetti, C. Gimond, J.-C. Chambard, T. Touboul, D. Roux, J. Pouysségur, and G. Pagès. Extracellular signal-regulated kinases phosphorylate mitogen-activated protein kinase phosphatase 3/DUSP6 at serines 159 and 197, two sites critical for its proteasomal degradation. *Mol Cell Biol*, 25(2):854–864, Jan 2005.

[103] N. I. Markevich, J. B. Hoek, and B. N. Kholodenko. Signaling switches and bistability arising from multisite phosphorylation in protein kinase cascades. *J Cell Biol*, 164(3):353–359, Feb 2004.

[104] J. L. Marks, Y. Gong, D. Chitale, B. Golas, M. D. McLellan, Y. Kasai, L. Ding, E. R. Mardis, R. K. Wilson, D. Solit, R. Levine, K. Michel, R. K. Thomas, V. W. Rusch, M. Ladanyi, and W. Pao. Novel MEK1 mutation identified by mutational analysis of epidermal growth factor receptor signaling pathway genes in lung adenocarcinoma. *Cancer Res*, 68(14):5524–5528, Jul 2008.

[105] M. M. McKay, D. A. Ritt, and D. K. Morrison. Signaling dynamics of the KSR1 scaffold complex. *Proc Natl Acad Sci U S A*, 106(27):11022–11027, Jul 2009.

[106] N. Meyer and L. Z. Penn. Reflecting on 25 years with MYC. *Nat Rev Cancer*, 8(12):976–990, Dec 2008.

[107] M. Miaczynska, L. Pelkmans, and M. Zerial. Not just a sink: endosomes in control of signal transduction. *Curr Opin Cell Biol*, 16(4):400–406, Aug 2004.

[108] T. A. Millward, S. Zolnierowicz, and B. A. Hemmings. Regulation of protein kinase cascades by protein phosphatase 2A. *Trends Biochem Sci*, 24(5):186–191, May 1999.

[109] S. Morton, R. J. Davis, A. McLaren, and P. Cohen. A reinvestigation of the multisite phosphorylation of the transcription factor c-Jun. *EMBO J*, 22(15):3876–3886, Aug 2003.

[110] N. K. Mukhopadhyay, D. J. Price, J. M. Kyriakis, S. Pelech, J. Sanghera, and J. Avruch. An array of insulin-activated, proline-directed serine/threonine protein kinases phosphorylate the p70 S6 kinase. *J Biol Chem*, 267(5):3325–3335, Feb 1992.

[111] J. E. Murphy, B. E. Padilla, B. Hasdemir, G. S. Cottrell, and N. W. Bunnett. Endosomes: a legitimate platform for the signaling train. *Proc Natl Acad Sci U S A*, 106(42):17615–17622, Oct 2009.

[112] L. O. Murphy, S. Smith, R.-H. Chen, D. C. Fingar, and J. Blenis. Molecular interpretation of ERK signal duration by immediate early gene products. *Nat Cell Biol*, 4(8):556–564, Aug 2002.

[113] T. Nakakuki, M. R. Birtwistle, Y. Saeki, N. Yumoto, K. Ide, T. Nagashima, L. Brusch, B. A. Ogunnaike, M. Okada-Hatakeyama, and B. N. Kholodenko. Ligand-specific c-Fos expression emerges from the spatiotemporal control of ErbB network dynamics. *Cell*, 141(5):884–896, May 2010.

[114] R. A. Nemenoff, S. Winitz, N. X. Qian, V. Van Putten, G. L. Johnson, and L. E. Heasley. Phosphorylation and activation of a high molecular weight form of phospholipase A2 by p42 microtubule-associated protein 2 kinase and protein kinase C. *J Biol Chem*, 268(3):1960–1964, Jan 1993.

[115] C. G. A. Network. Comprehensive molecular characterization of human colon and rectal cancer. *Nature*, 487(7407):330–337, Jul 2012.

[116] R. A. O'Neill, A. Bhamidipati, X. Bi, D. Deb-Basu, L. Cahill, J. Ferrante, E. Gentalen, M. Glazer, J. Gossett, K. Hacker, C. Kirby, J. Knittle, R. Loder, C. Mastroieni, M. Maclaren, T. Mills, U. Nguyen, N. Parker, A. Rice, D. Roach, D. Suich, D. Voehringer, K. Voss, J. Yang, T. Yang, and P. B. Vander Horn. Isoelectric focusing technology quantifies protein signaling in 25 cells. *Proc Natl Acad Sci U S A*, 103(44):16153–16158, Oct 2006.

[117] A. Ooi, T. Takehana, X. Li, S. Suzuki, K. Kunitomo, H. Iino, H. Fujii, Y. Takeda, and Y. Dobashi. Protein overexpression and gene amplification of HER-2 and EGFR in colorectal cancers: an immunohistochemical and fluorescent in situ hybridization study. *Mod Pathol*, 17(8):895–904, Aug 2004.

[118] G. Pagès, P. Lenormand, G. L'Allemain, J. C. Chambard, S. Meloche, and J. Pouysségur. Mitogen-activated protein kinases p42mapk and p44mapk are required for fibroblast proliferation. *Proc Natl Acad Sci U S A*, 90(18):8319–8323, Sep 1993.

[119] D. R. Parks, M. Roederer, and W. A. Moore. A new "Logicle" display method avoids deceptive effects of logarithmic scaling for low signals and compensated data. *Cytometry Part A*, 69A(6):541–551, Jun 2006.

[120] K. I. Patterson, T. Brummer, P. M. O'Brien, and R. J. Daly. Dual-specificity phosphatases: critical regulators with diverse cellular targets. *Biochem J*, 418(3):475–489, Mar 2009.

[121] G. Pearson, F. Robinson, T. Beers Gibson, B. E. Xu, M. Karandikar, K. Berman, and M. H. Cobb. Mitogen-activated protein (MAP) kinase pathways: regulation and physiological functions. *Endocr Rev*, 22(2):153–183, Apr 2001.

[122] R. M. Perrett, R. C. Fowkes, C. J. Caunt, K. Tsaneva-Atanasova, C. G. Bowsher, and C. A. McArdle. Signaling to extracellular signal-regulated kinase from ErbB1 kinase and protein kinase C: feedback, heterogeneity, and gating. *J Biol Chem*, 288(29):21001–21014, Jul 2013.

[123] S. Persad, S. Attwell, V. Gray, N. Mawji, J. T. Deng, D. Leung, J. Yan, J. Sanghera, M. P. Walsh, and S. Dedhar. Regulation of protein kinase B/Akt-serine 473 phosphorylation by integrin-linked kinase: critical roles for kinase activity and amino acids arginine 211 and serine 343. *J Biol Chem*, 276(29):27462–27469, Jul 2001.

[124] S. Prabakaran, R. A. Everley, I. Landrieu, J.-M. Wieruszeski, G. Lippens, H. Steen, and J. Gunawardena. Comparative analysis of Erk phosphorylation suggests a mixed strategy for measuring phospho-form distributions. *Mol Syst Biol*, 7:482, Apr 2011.

[125] N. Pullen, P. B. Dennis, M. Andjelkovic, A. Dufner, S. C. Kozma, B. A. Hemmings, and G. Thomas. Phosphorylation and activation of p70s6k by PDK1. *Science*, 279(5351):707–710, Jan 1998.

[126] A. Raue, M. Schilling, J. Bachmann, A. Matteson, M. Schelker, M. Schelke, D. Kaschek, S. Hug, C. Kreutz, B. D. Harms, F. J. Theis, U. Klingmüller, and J. Timmer. Lessons learned from quantitative dynamical modeling in systems biology. *PLoS One*, 8(9):e74335, Sep 2013.

[127] S. M. Reppert and D. R. Weaver. Coordination of circadian timing in mammals. *Nature*, 418(6901):935–941, Aug 2002.

[128] F. J. Richards. A flexible growth function for empirical use. *Journal of Experimental Botany*, 10(2):290–301, 1959.

[129] R. Roskoski, Jr. RAF protein-serine/threonine kinases: structure and regulation. *Biochem Biophys Res Commun*, 399(3):313–317, Aug 2010.

[130] R. Roskoski, Jr. The ErbB/HER family of protein-tyrosine kinases and cancer. *Pharmacol Res*, 79:34–74, Jan 2014.

[131] C. Salazar and T. Höfer. Kinetic models of phosphorylation cycles: a systematic approach using the rapid-equilibrium approximation for protein-protein interactions. *Biosystems*, 83(2-3):195–206, Mar 2006.

[132] C. Salazar and T. Höfer. Multisite protein phosphorylation - from molecular mechanisms to kinetic models. *FEBS J*, 276(12):3177–3198, Jun 2009.

[133] S. D. M. Santos, P. J. Verveer, and P. I. H. Bastiaens. Growth factor-induced MAPK network topology shapes Erk response determining PC-12 cell fate. *Nat Cell Biol*, 9(3):324–330, Mar 2007.

[134] M. Schilling, T. Maiwald, S. Hengl, D. Winter, C. Kreutz, W. Kolch, W. D. Lehmann, J. Timmer, and U. Klingmüller. Theoretical and experimental analysis links isoform-specific ERK signalling to cell fate decisions. *Mol Syst Biol*, 5(1):334, Jan 2009.

[135] M. Schwab. Amplification of oncogenes in human cancer cells. *Bioessays*, 20(6):473–479, Jun 1998.

[136] H. Shankaran, D. L. Ippolito, W. B. Chrisler, H. Resat, N. Bollinger, L. K. Opresko, and H. S. Wiley. Rapid and sustained nuclear-cytoplasmic ERK oscillations induced by epidermal growth factor. *Mol Syst Biol*, 5(1):332, Jan 2009.

[137] S. K. Shenoy and R. J. Lefkowitz. β-arrestin-mediated receptor trafficking and signal transduction. *Trends Pharmacol Sci*, 32(9):521–533, Sep 2011.

[138] S.-Y. Shin, O. Rath, S.-M. Choo, F. Fee, B. McFerran, W. Kolch, and K.-H. Cho. Positive- and negative-feedback regulations coordinate the dynamic behavior of the Ras-Raf-MEK-ERK signal transduction pathway. *J Cell Sci*, 122(Pt 3):425–435, Feb 2009.

[139] G. Shinar and M. Feinberg. Structural sources of robustness in biochemical reaction networks. *Science*, 327(5971):1389–1391, Mar 2010.

[140] A. Sigal, R. Milo, A. Cohen, N. Geva-Zatorsky, Y. Klein, Y. Liron, N. Rosenfeld, T. Danon, N. Perzov, and U. Alon. Variability and memory of protein levels in human cells. *Nature*, 444(7119):643–646, Nov 2006.

[141] D. J. Slamon, W. Godolphin, L. A. Jones, J. A. Holt, S. G. Wong, D. E. Keith, W. J. Levin, S. G. Stuart, J. Udove, and A. Ullrich. Studies of the HER-2/neu proto-oncogene in human breast and ovarian cancer. *Science*, 244(4905):707–712, May 1989.

[142] J. A. Smith, C. E. Poteet-Smith, K. Malarkey, and T. W. Sturgill. Identification of an extracellular signal-regulated kinase (ERK) docking site in ribosomal S6 kinase, a sequence critical for activation by ERK in vivo. *J Biol Chem*, 274(5):2893–2898, Jan 1999.

[143] D. B. Solit, L. A. Garraway, C. A. Pratilas, A. Sawai, G. Getz, A. Basso, Q. Ye, J. M. Lobo, Y. She, I. Osman, T. R. Golub, J. Sebolt-Leopold,

W. R. Sellers, and N. Rosen. BRAF mutation predicts sensitivity to MEK inhibition. *Nature*, 439(7074):358–362, Jan 2006.

[144] K. Solly, X. Wang, X. Xu, B. Strulovici, and W. Zheng. Application of real-time cell electronic sensing (RT-CES) technology to cell-based assays. *Assay Drug Dev Technol*, 2(4):363–372, Aug 2004.

[145] L. P. Sousa, I. Lax, H. Shen, S. M. Ferguson, P. De Camilli, and J. Schlessinger. Suppression of EGFR endocytosis by dynamin depletion reveals that EGFR signaling occurs primarily at the plasma membrane. *Proc Natl Acad Sci U S A*, 109(12):4419–4424, Mar 2012.

[146] R. Straube. Sensitivity and robustness in covalent modification cycles with a bifunctional converter enzyme. *Biophys J*, 105(8):1925–1933, Oct 2013.

[147] T. W. Sturgill, L. B. Ray, E. Erikson, and J. L. Maller. Insulin-stimulated MAP-2 kinase phosphorylates and activates ribosomal protein S6 kinase ii. *Nature*, 334(6184):715–718, Aug 1988.

[148] O. E. Sturm, R. Orton, J. Grindlay, M. Birtwistle, V. Vyshemirsky, D. Gilbert, M. Calder, A. Pitt, B. Kholodenko, and W. Kolch. The mammalian MAPK/ERK pathway exhibits properties of a negative feedback amplifier. *Sci Signal*, 3(153):ra90, Dec 2010.

[149] Y. Tanaka, T. Ogasawara, Y. Asawa, H. Yamaoka, S. Nishizawa, Y. Mori, T. Takato, and K. Hoshi. Growth factor contents of autologous human sera prepared by different production methods and their biological effects on chondrocytes. *Cell Biol Int*, 32(5):505–514, May 2008.

[150] Z. Tang, S. Dai, Y. He, R. A. Doty, L. D. Shultz, S. B. Sampson, and C. Dai. MEK guards proteome stability and inhibits tumor-suppressive amyloidogenesis via HSF1. *Cell*, 160(4):729–744, Feb 2015.

[151] T. Tanoue and E. Nishida. Molecular recognitions in the MAP kinase cascades. *Cell Signal*, 15(5):455–462, May 2003.

[152] C. G. A. R. N. TCGA. Comprehensive molecular characterization of urothelial bladder carcinoma. *Nature*, 507(7492):315–322, Mar 2014.

[153] T. Toni, Y.-i. Ozaki, P. Kirk, S. Kuroda, and M. P. H. Stumpf. Elucidating the in vivo phosphorylation dynamics of the ERK MAP kinase using

quantitative proteomics data and bayesian model selection. *Mol Biosyst*, 8(7):1921–1929, Jul 2012.

[154] S. Torii, M. Kusakabe, T. Yamamoto, M. Maekawa, and E. Nishida. Sef is a spatial regulator for Ras/MAP kinase signaling. *Dev Cell*, 7(1):33–44, Jul 2004.

[155] C. E. Turner. Paxillin interactions. *J Cell Sci*, 113 Pt 23:4139–4140, Dec 2000.

[156] S. Uda, T. H. Saito, T. Kudo, T. Kokaji, T. Tsuchiya, H. Kubota, Y. Komori, Y.-i. Ozaki, and S. Kuroda. Robustness and compensation of information transmission of signaling pathways. *Science*, 341(6145):558–561, Aug 2013.

[157] R. R. Vaillancourt, L. E. Heasley, J. Zamarripa, B. Storey, M. Valius, A. Kazlauskas, and G. L. Johnson. Mitogen-activated protein kinase activation is insufficient for growth factor receptor-mediated PC12 cell differentiation. *Mol Cell Biol*, 15(7):3644–3653, Jul 1995.

[158] A. C. Ventura, J.-A. Sepulchre, and S. D. Merajver. A hidden feedback in signaling cascades is revealed. *PLoS Comput Biol*, 4(3):e1000041, Mar 2008.

[159] G. von Dassow, E. Meir, E. M. Munro, and G. M. Odell. The segment polarity network is a robust developmental module. *Nature*, 406(6792):188–192, Jul 2000.

[160] D. von Helversen. Gesang des Männchens und Lautschema des Weibchens bei der Feldheuschrecke Chorthippus biguttulus (Orthoptera, Acrididae). *Journal of comparative physiology*, 81(4):381–422, 1972.

[161] T. Wada and J. M. Penninger. Mitogen-activated protein kinases in apoptosis regulation. *Oncogene*, 23(16):2838–2849, Apr 2004.

[162] A. Wells, J. B. Welsh, C. S. Lazar, H. S. Wiley, G. N. Gill, and M. G. Rosenfeld. Ligand-induced transformation by a noninternalizing epidermal growth factor receptor. *Science*, 247(4945):962–964, Feb 1990.

[163] R. H. Whitehead, F. A. Macrae, D. J. St John, and J. Ma. A colon cancer cell line (LIM1215) derived from a patient with inherited nonpolyposis colorectal cancer. *J Natl Cancer Inst*, 74(4):759–765, Apr 1985.

[164] F. Witzel, R. Fritsche-Guenther, N. Lehmann, A. Sieber, and N. Blüth-gen. Analysis of impedance-based cellular growth assays. *Bioinformatics*, 31(16):2705–2712, Apr 2015.

[165] F. Witzel, L. Maddison, and N. Blüthgen. How scaffolds shape MAPK signaling: what we know and opportunities for systems approaches. *Front Physiol*, 3:475, Dec 2012.

[166] W. Wunderlich, I. Fialka, D. Teis, A. Alpi, A. Pfeifer, R. G. Parton, F. Lottspeich, and L. A. Huber. A novel 14-kilodalton protein interacts with the mitogen-activated protein kinase scaffold mp1 on a late endoso-mal/lysosomal compartment. *J Cell Biol*, 152(4):765–776, Feb 2001.

[167] T. Yamamoto, M. Ebisuya, F. Ashida, K. Okamoto, S. Yonehara, and E. Nishida. Continuous ERK activation downregulates antiproliferative genes throughout G1 phase to allow cell-cycle progression. *Curr Biol*, 16(12):1171–1182, Jun 2006.

[168] J. Yang and W. S. Hlavacek. Scaffold-mediated nucleation of protein sig-naling complexes: elementary principles. *Math Biosci*, 232(2):164–173, Aug 2011.

[169] J. J. Yeh, E. D. Routh, T. Rubinas, J. Peacock, T. D. Martin, X. J. Shen, R. S. Sandler, H. J. Kim, T. O. Keku, and C. J. Der. KRAS/BRAF mu-tation status and ERK1/2 activation as biomarkers for MEK1/2 inhibitor therapy in colorectal cancer. *Mol Cancer Ther*, 8(4):834–843, Apr 2009.

[170] T. C. Yeh, V. Marsh, B. A. Bernat, J. Ballard, H. Colwell, R. J. Evans, J. Parry, D. Smith, B. J. Brandhuber, S. Gross, A. Marlow, B. Hurley, J. Lyssikatos, P. A. Lee, J. D. Winkler, K. Koch, and E. Wallace. Biological characterization of ARRY-142886 (AZD6244), a potent, highly selective mitogen-activated protein kinase kinase 1/2 inhibitor. *Clin Cancer Res*, 13(5):1576–1583, Mar 2007.

[171] T. M. Yi, Y. Huang, M. I. Simon, and J. Doyle. Robust perfect adaptation in bacterial chemotaxis through integral feedback control. *Proc Natl Acad Sci U S A*, 97(9):4649–4653, Apr 2000.

[172] S. Yoon and R. Seger. The extracellular signal-regulated kinase: multiple substrates regulate diverse cellular functions. *Growth Factors*, 24(1):21–44, Mar 2006.

[173] W. Yu, W. J. Fantl, G. Harrowe, and L. T. Williams. Regulation of the MAP kinase pathway by mammalian Ksr through direct interaction with MEK and ERK. *Curr Biol*, 8(1):56–64, Jan 1998.

[174] E. E. Zhang and S. A. Kay. Clocks not winding down: unravelling circadian networks. *Nat Rev Mol Cell Biol*, 11(11):764–776, Nov 2010.

[175] Y. Zhang, Z. Dong, M. Nomura, S. Zhong, N. Chen, A. M. Bode, and Z. Dong. Signal transduction pathways involved in phosphorylation and activation of p70S6K following exposure to UVA irradiation. *J Biol Chem*, 276(24):20913–20923, Jun 2001.

[176] Y. Zhao and Z. Y. Zhang. The mechanism of dephosphorylation of extracellular signal-regulated kinase 2 by mitogen-activated protein kinase phosphatase 3. *J Biol Chem*, 276(34):32382–32391, Aug 2001.

[177] X.-M. Zhu, L. Yin, L. Hood, and P. Ao. Calculating biological behaviors of epigenetic states in the phage lambda life cycle. *Funct Integr Genomics*, 4(3):188–195, Jul 2004.

Acknowledgements

Mein erster Dank gilt meinem Doktorvater Nils Blüthgen, der mir keine bessere Betreuung hätte zukommen lassen können. Er war stets für Gespräche verfügbar und war eine verlässliche Quelle fuer neue Ideen, Inspiration und vorallem Motivation. Seine Begeisterung für die Wissenschaft und seine sympathische Natur waren auch der Klebstoff für das leichte Entstehen von interessanten Kooperationen. An dieser Stelle möchte ich mich auch bei Oliver Ebenhöh bedanken, der mit seiner großartigen Vorlesung und seiner tollen Betreuung meiner Diplomarbeit die Weichen gestellt hat für meinen Einstieg in die Theoretische Biologie. Auch war er nicht unbeteiligt daran, dass ich meine Stelle bei Nils bekommen habe, wie sich im Nachhinein herausgestellt hat.

Einen herzlichen Dank an Hanspeter Herzel, der sich die Zeit dafür genommen hat, meine Arbeit zu begutachten und auch für inhaltliche sowie formale Fragen zum Thema Dissertation immer ein Ohr hatte. Ebenfalls einen großen Dank an meinen Gutachter Ronny Straube, den ich auf einer Konferenz kennengelernt habe und welcher als Experte der Phosphorylierungskinetik spontan eingewilligt hat, meine Arbeit kritisch unter die Lupe zu nehmen. Ich danke Peter Hammerstein, der den Vorsitz meiner Promotionskommission übernommen hat und welchen meine Verteidigung ähnlich in Aufregung versetzt hat wie mich selbst. Danke für dein Mitfiebern und deine Unterstützung Peter! Herzlichen Dank auch an Britta Eickholt, die auch wesentlich zu der interessanten und sympathischen Atmosphäre meiner Verteidigung beigetragen hat.

Lieben Dank an die Mitglieder des inoffiziellen ITB Chores, Elvira Lauterbach, Agnes Rosahl, Paula Kuokkanen, José Donoso, Jorge Jaramillo, Fridolin Groß und Torsten Groß, durch euch habe ich mich am ITB nach all den Weihnachtskonzerten richtig heimisch gefühlt. WOW zu der Überraschungseinlage zu meiner Verteidigung, ihr habt für einen unvergesslichen Tag gesorgt und mich sehr gerührt!

Danke für den netten und kompetenten administrativen support von Elvira, Andreas Hantschmann, Jana Lahmer und auch Pascal Schulthess, den ich oft und gern von der Arbeit abgehalten habe, wenn mein Mac nicht so wollte wie ich. Meinen Langzeit-Bürokollegen Bertram Klinger und Pascal Schulthess bin ich dankbar für die lustige Atmosphäre und die gegenseitige Unterstützung und Motivation. Danke an Grigory Bordyugov und Max Schelker für ihren input zum Fitten der Lösungen von Differentialgleichungen an experimentelle Daten.

Ein besonderer Dank an Johannes Meisig - gemeinsam haben wir die Operation „finish line" in Angriff genommen und regelmäßig über unsere Arbeit diskutiert. Danke Johannes für deine hilfreichen Ratschläge und deine Freundschaft!

Neben dem ITB habe ich auch viel Zeit in den Laboren der Molekularen Tumorpathologie verbracht. Herzlichen Dank an Anja Sieber und Raphaela Fritsche-Guenther, die mir alles Notwendige beigebracht haben, um meine eigenen Experimente zu planen und durchzuführen. Danke an Markus Morkel für die Einführung in die hohe Kunst des Klonierens. Danke an Sandra Schrötter für die gemeinsam angefertigten Experimente am Charité CCO. Herzlichen Dank auch an Cornelia Gieseler, Kerstin Möhr, Judith Seidemann, Torben Redmer, Kathleen Klotz-Noack, Pamela Riemer, Felix Bormann, Natalia Kuhn, Paula Medina-Perez und Christina Kuznia, die mir stets bei kleinen und großen Problemen im Labor geholfen haben. Ein besonderes Dankeschön an Nadine Lehmann für die tolle Zusammenarbeit. Ohne deine Unterstützung Nadine, hätte meine Promotion mindestens noch ein Jahr länger gedauert!

Zu guter letzt ein riesiges Dankeschön an meinen Bruder Thomas und meine Freunde, insbesondere Grisha, Margit, Luis und Sarah, für ihren Glauben an mich und ihre Unterstützung.